Earl Stanhope, Charles Wildbore

On Spherical Motion. By the Rev. Charles Wildbore; Communicated by Earl Stanhope, F. R. S.

Earl Stanhope, Charles Wildbore

On Spherical Motion. By the Rev. Charles Wildbore; Communicated by Earl Stanhope, F. R. S.

ISBN/EAN: 9783337102333

Printed in Europe, USA, Canada, Australia, Japan

Cover: Foto ©berggeist007 / pixelio.de

More available books at **www.hansebooks.com**

XXIV. *On Spherical Motion. By the Rev.* Charles Wildbore; *communicated by Earl* Stanhope, *F. R. S.*

Read June 24, 1790.

THIS Paper, which has coſt me much pains in patient inveſtigation, is occaſioned by that of Mr. LANDEN, in the Philoſophical Tranſactions, Vol. LXXV. Part II. I am no ſtranger to this gentleman's great judgement and abilities in theſe abſtruſe ſpeculations, but have a very high opinion of both; yet I could not but think it ſtrange, that two ſuch mathematicians as M. D'ALEMBERT and M. L. EULER ſhould both follow one another on the ſame ſubject, both agree, and ſtill not be right. I therefore reſolved to try to dive to the bottom of their ſolutions, which thoſe who are acquainted with the ſubject know to be no light taſk; and, if poſſible, to give the ſolution, independent of the perplexing conſideration of a momentary axis changing its place both in the body and in abſolute ſpace every inſtant; and which I look upon as not abſolutely eſſential to the determination of the body's motion. But finding that I could not thus ſo readily ſhew the agreement or diſagreement of my concluſions with thoſe of the gentlemen who have preceded me in this enquiry; I have alſo added the inveſtigation of the properties of this axis. And I ſuppoſe it will be found, that I have added many properties unknown before, or at leaſt unnoticed by any of them.

M. LANDEN'S

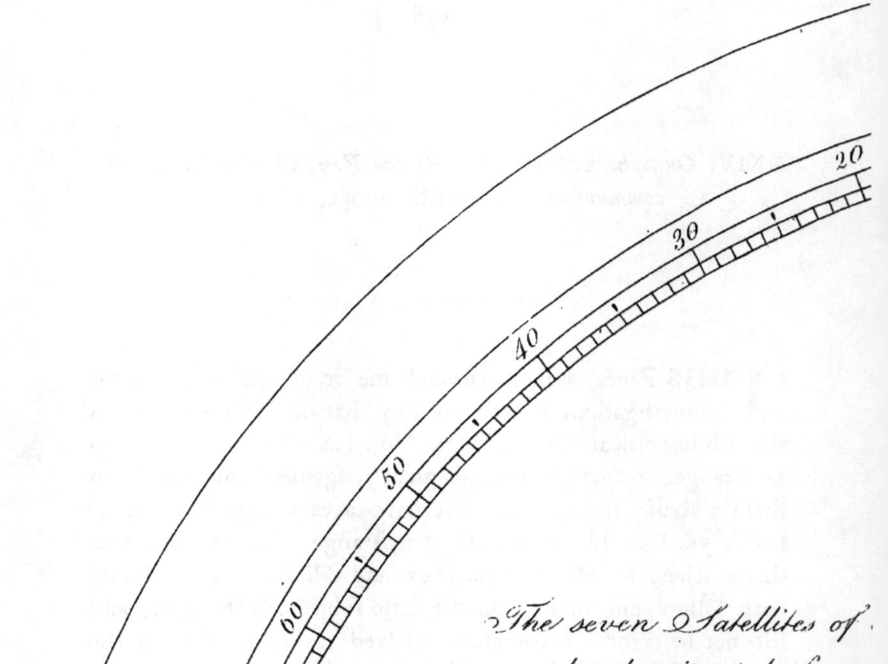
20
30
40
50
60
70
80
The seven Satellites of
appeared to be situated the 18:
at 7.h 51'. 54." 1789.

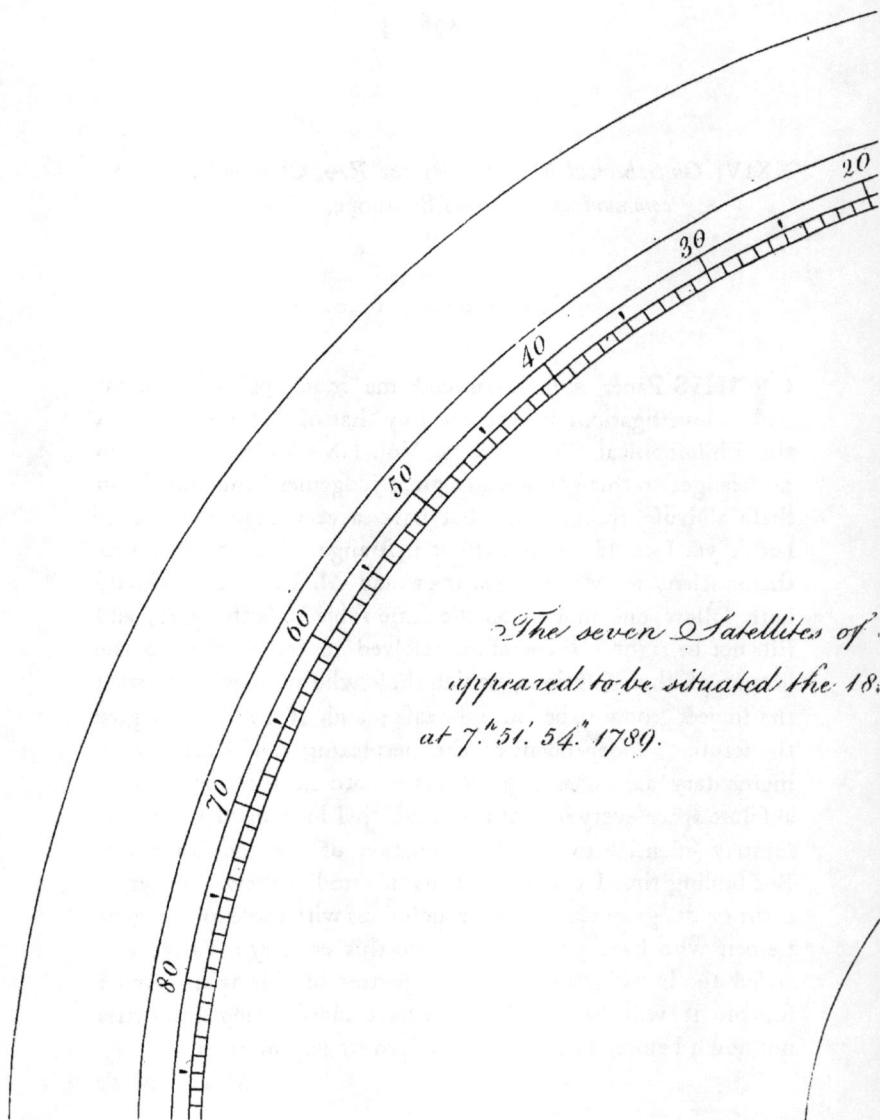

of Saturn as they

18th of Octr

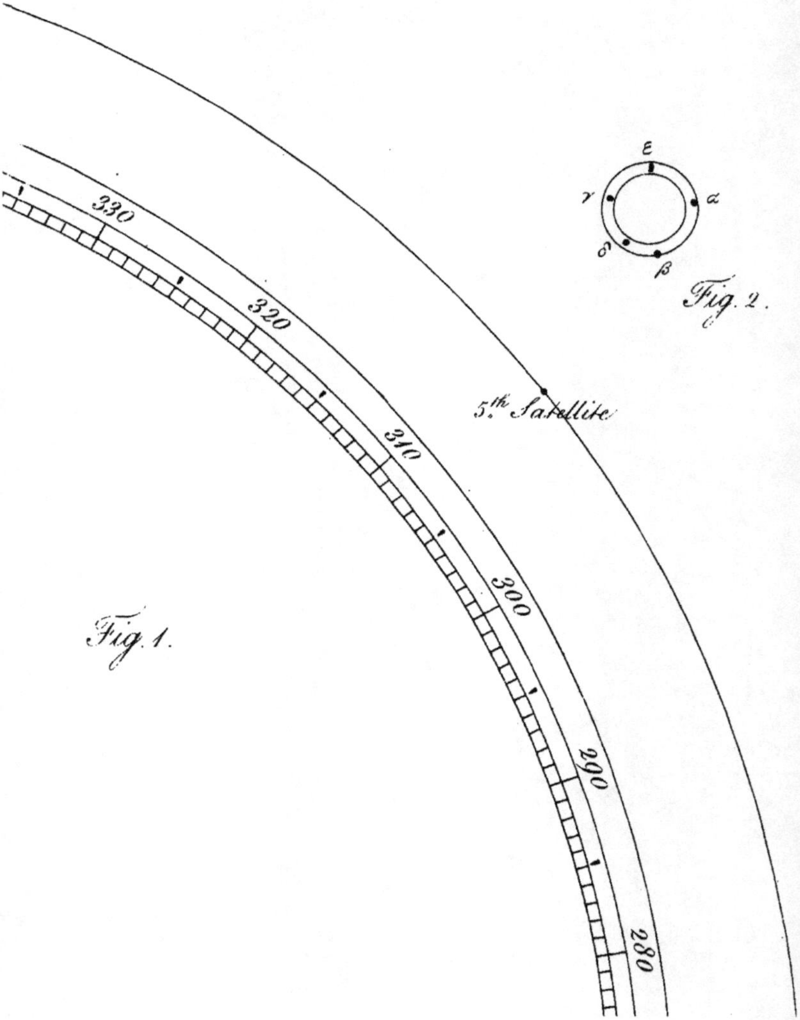

Fig. 2.

Fig. 1.

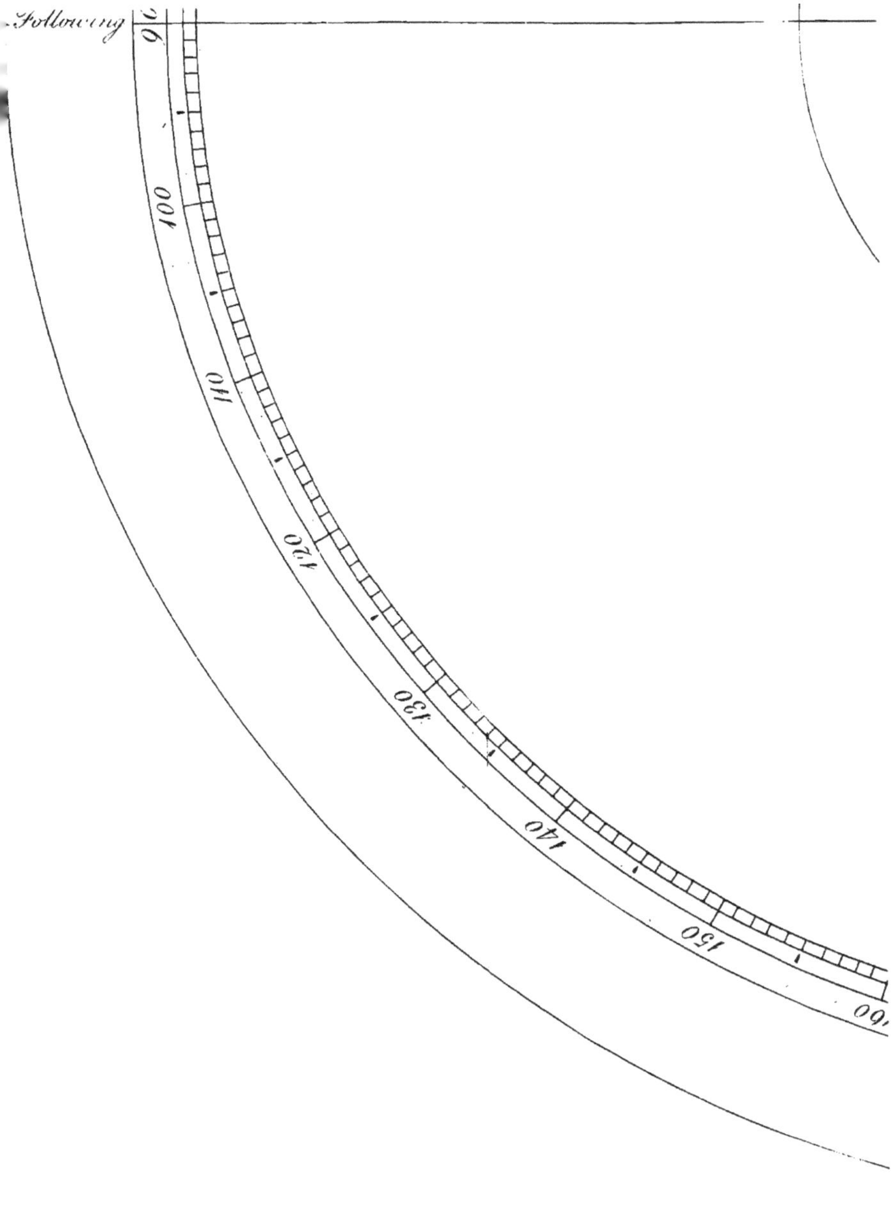
Following
100
110
120
130
140
150

4.th Satellite.

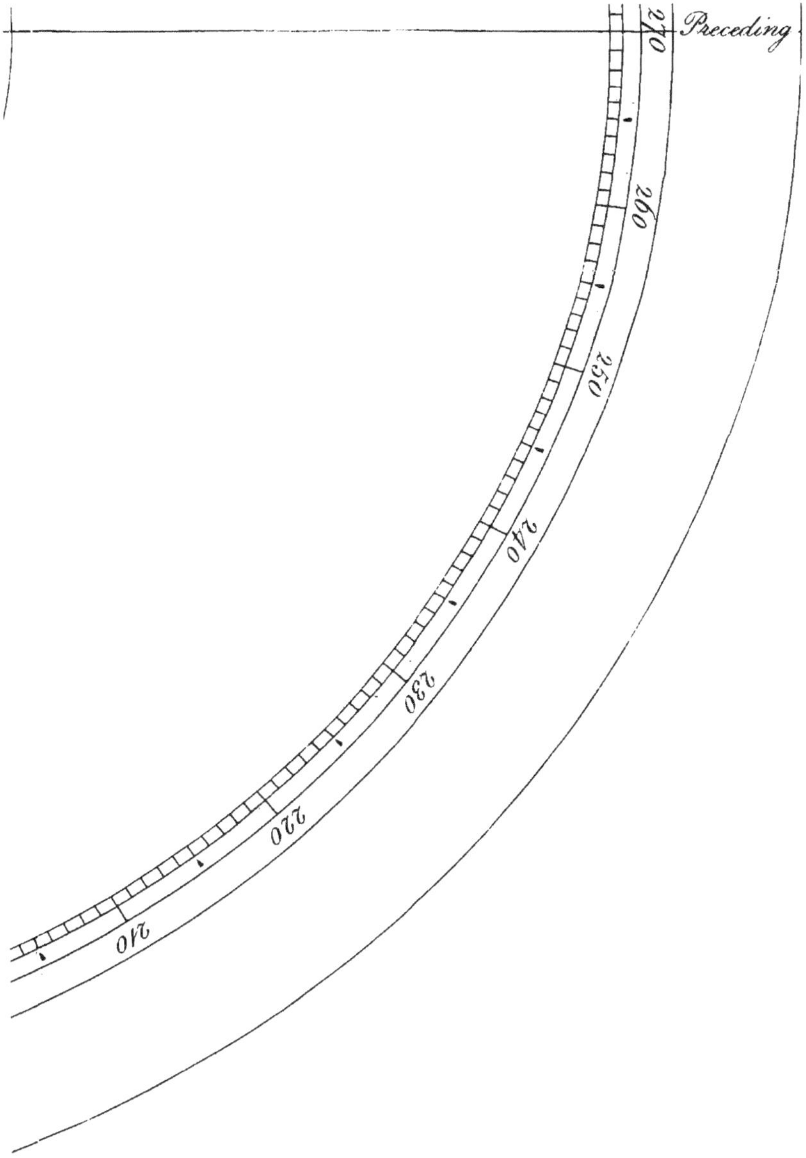
Preceding
270
260
250
240
230
220
210

M. LANDEN's very important diſcovery, that every body, be its form ever ſo irregular, will revolve in the ſame manner as if its maſs were equally divided and placed in the eight angles, or diſpoſed in the eight octants of a regular parallelopipedon, whoſe moments of *inertia* round its three permanent axes are the ſame as thoſe of the body, ſerves admirably to ſhorten the inveſtigation, and render the ſolution perſpicuous. I have therefore here taken its truth for granted, becauſe it is alſo exactly agreeable to the ſolutions of the other gentlemen, and ſaves the trouble of repeating what they have done before. I have alſo ſhewn wherein, and why, his ſolution differs from theirs, and proved, as I think, undeniably, in what reſpects it is defective.

That the *inertia*, or, as M. EULER calls it, the *momentum* of *inertia*, is equal to the fluent or ſum of every particle of the body drawn into the ſquare of its diſtance from the axis of motion; and the determination of the three permanent axes, or the demonſtration that there are, at leaſt, three ſuch axes in every body, round any one of which, if it revolved, the velocity would be for ever uniform, I have alſo taken for granted, becauſe theſe things have been proved before, and all the gentlemen are agreed in them. Difficulties that occurred I have not concealed, but ſhewn how to obviate, and endeavoured to place the truth in as clear a light as poſſible; which to diſcover is my wiſh, or to welcome it by whomſoever found.

PROPOSITION I.

Whilſt a globe, whoſe centre is at reſt, revolves with a given velocity about an axis paſſing through that centre, to

 find

find with what velocity any great circle on the ſurface, but oblique to that axis, moves along itſelf.

Let I (Tab. XX. fig. 1.) be the centre, and BL*b* the axis round which the globe revolves with a velocity = *c* meaſured along the great circle GH, whoſe plane is perpendicular to that axis, and HSG*s* any great circle whoſe plane is oblique to the axis, ESF and *esf* two leſſer circles of the ſphere parallel to the great circle GH, and touching HSG*s* in S and *s*; then, as the radius BI which may be ſuppoſed unity : *c* :: the radius of the leſs circle ESF = the ſine of the arc BE or BS : the velocity along the circle ESF = the abſolute velocity of the point S on the ſurface of the globe: but the point S is alſo upon the great circle GSH*s*, and therefore this is alſo equal to the velocity of the point *s* along the great circle GSH*s*; and for the ſame reaſons the point S, which is diametrically oppoſite to S on the ſurface, has alſo the ſame velocity. Let P be any other point in the great circle GSH*s*; then, ſince as the globe revolves the diſtances SP and *s*P always continue invariable, the velocity of the point P in the circle HPS in the direction of the periphery of the circle itſelf muſt be equal to that of S and *s*; and is therefore the velocity of every point of this circle along its own periphery.

Corollary 1. Hence it follows, that in whatſoever manner a globe revolves, its velocity meaſured on the ſame great circle on its ſurface muſt be the ſame at the ſame time at every point of the periphery of that circle.

Corollary 2. Conſequently, howſoever the plane of a great circle varies its motion, the velocity at any inſtant is at every point of the periphery equal along its own plane.

DEFI-

DEFINITION.

The points S and *s*, where a great circle from the poles B and *b* of the natural axis cuts any great circle GSH*s* (at right angles) I call the nodes of that great circle.

Corollary 3. If O be the pole of the great circle HSG*s*, then the globe may be confidered as moving round the axis whofe pole is O with a velocity $= c \times \frac{\text{fine of BS}}{\text{BI}}$, whilft the pole O is carried along the leffer circle AOA, which is parallel to the midcircle GH with a velocity $= c \times \frac{\text{fin. O}b}{\text{BI}} = c \times \frac{\text{cof. BS}}{\text{BI}}$; and this way of confidering the motion, which is ufeful in what follows, comes to the very fame as the motion along the great or midcircle GH with the velocity $= c$, becaufe $c^2 \times \frac{\text{f. BS}^2}{\text{BI}^2} + c^2 \times \frac{\text{cof. BS}^2}{\text{BI}^2} = c^2$. Confequently, the fum of the fquares of the velocities at the node and pole of any great circle upon a fpherical furface thus revolving, is equal to the fquare of the velocity round the natural (or momentary) axis BI*b*.

Corollary 4. Since the pole O is at 90° diftance from the node S, its motion can have no effect at S or *s*, the motion at the nodes, therefore, of the great circle HSG*s* is that of the great circle along its own proper plane; but any other point, as P, partakes both of the motion along the circle, and the motion of its pole. The direction of its motion being along the leffer circle P*p*, parallel to FSE, and its velocity therein $= c \times \frac{\text{f. BP}}{\text{f. BS}}$; the velocity of P therefore, in the direction of the great circle OP, which is perpendicular to SP in P, is $= c\sqrt{\left(\frac{\text{f. BP}^2}{\text{f. BS}^2} - \frac{\text{f. BS}^2}{\text{BI}^2}\right)}$, and along the great circle BP its velocity $= 0$.

 PROPO-

PROPOSITION II.

Suppofing the centre of a fphere to be at reft, whilft the furface moves round it in any manner whatfoever; then, if the fame invariable point O, confidered as the pole of an axis of the fphere, be itfelf in motion, the angular velocity of the fpherical furface about that axis will be unequable, or that of one point therein different from that of another.

For, let I (fig. 2.) be the centre of the fphere; draw the great circle POF perpendicular to the direction of the motion of the furface at O; then muft the pole of this motion neceffarily be in fome point P of this great circle POF. Let FC be the great circle whofe pole is P, and LQ that whofe pole is O; then, the velocity of any point F of the great circle FC muft, by the preceding propofition, be equal to that of any other point H thereof. Let that velocity be reprefented by the equal arches FG and HK, and from the pole O draw the great circles OGM, OHN, and OKA, cutting the great circle LQ in M, N, and A; then muft LM reprefent the angular velocity of the point F about the axis IO, and NA that of the point H. But, by Prop. 9. Lib. III. THEODOSII Sphericorum, LM is greater than NA; and confequently the angular velocity of the point F about IO is greater than that of H; and confequently the angular velocity of the furface about the axis IO is unequable.

Corollary. Hence, about whatever axis the angular motion of a fphere is equable, the pole of that axis, and confequently the axis itfelf, muft be at reft at the inftant. Different motions may have different correfpondent poles, and confequently, when the motion is variable, the place of the pole of equable motion

motion on the ſurface may vary; but whatever point on the ſurface correſponds with that pole muſt at the inſtant be at reſt.

PROPOSITION III.

Let ABC (fig. 3.) be an octant of a ſpherical ſurface in motion, while the centre is at reſt; and let the velocity of the great circle BC in its own plane $=a$, and in a ſenſe from B towards C; that of CA in the ſenſe from C towards A $=b$, and of AB from A towards B $=c$. If theſe three velocities a, b, and c, be conſtant, the ſpherical ſurface will always revolve uniformly about the ſame axis of the ſphere at reſt in abſolute ſpace.

For, let ABC, abc, be two poſitions of the revolving octant indefinitely near each other, Aa, Bb, and Cc, the tracks of A, B, and C, in abſolute ſpace. Perpendicular to Aa draw the great circle SOA, and perpendicular to Bb the great circle BOQ, cutting SOA in O and CA in Q; then, becauſe Aa is indefinitely ſmall, the two triangles Apa right-angled at p, and a'Aq right-angled at A may be conſidered as plane ones, and are therefore ſimilar; and ſince the angles pAQ and qAa are both right ones, taking away qAp, which is common, the angles pAa, qAQ, muſt be equal; but as pA : pa :: c : b, likewiſe pA : pa :: ſ. paA : ſ. pAa, and paA $=p$Aq, pAa $=q$AQ; conſequently, as ſ. pAq : ſ. qAQ :: c : b, that is, the ſines of the angles BAS and CAS are proportional to the velocities along AB and CA; conſequently, the ſines of the arches SB and SC which are the meaſures of thoſe angles muſt be in the ſame ratio. In like manner it appears, that as ſ. CQ : ſ. AQ ::

$a : c ::$ ſ. CBQ : ſ. ABQ. Moreover, ſ. SOB : radius :: ſ. SB : ſ. BO :: ſ. AQ : ſ. AO. Through C and O draw the great circle COR; then, as ſ. AO : radius :: ſ. OAR : ſ. OR :: ſ. OAQ : ſ. OQ, or ſ. OR : ſ. OQ :: ſ. OAR : ſ. OAQ :: $c : b$, and for a like reaſon, ſ. OR : ſ. OS :: ſ. OBR : ſ. OBS :: $c : a$, that is, ſ. OR : c :: ſ. OQ : b :: ſ. OS : a, or ſ. OQ : ſ. OS :: $b : a$; but ſ. OQ : ſ. OS :: ſ. OCQ = ſ. AR : ſ. OCS = ſ. BR, or ſ. AR : ſ. BR :: $b : a$. Now, bc is ultimately perpendicular to AC in d, ſo the triangle Cdc being right-angled at d, the ſum of the angles Ccd, cCd muſt be = a right one, and their ſines are in the ratio of Cd : cd, or of $b : a$; but the ſum of the angles OCQ, OCS, is alſo a right one, and their ſines alſo have been proved to be in the ſame ratio of $b : a$, conſequently the angle OCQ = Ccd, and OCS = cCd, to OCS and cCd add the common angle OCQ, and the angle OCc muſt be = BCQ a right one: conſequently OC is perpendicular to Cc the track of the point C, as OA is, by hypotheſis, to Aa, and OB to Bb. The ſines of SO, QO, and RO, are as a, b, and c, alſo ſ. SO^2 + ſ. QO^2 + ſ. RO^2 by trigonometry = the ſquare of radius = 1; hence ſ. SO^2 + ſ. QO^2 = 1 − ſ. RO^2 = ſ. CO^2; ſ. SO^2 + ſ. RO^2 = ſ. BO^2, ſ. QO^2 + ſ. RO^2 = ſ. AO^2; conſequently, ſ. AO^2, ſ. BO^2 and ſ. CO^2 are as b^2+c^2, a^2+c^2, and a^2+b^2, or as Aa^2, Bb^2, and Cc^2; wherefore the velocities $\sqrt{b^2+c^2}$, $\sqrt{a^2+c^2}$, and $\sqrt{a^2+b^2}$, of the points A, B, C, are in directions perpendicular to AO, BO, and CO, and in the ratio of the ſines of the arches AO, BO, and CO, that is of the diſtances of the points A, B, and C, from the axis whoſe pole is O, the tracks of theſe points are therefore circles of the ſphere whoſe radii are thoſe diſtances. And ſo long as the velocities a, b, and c, are invariable, the points Q, R, and S, which are always at the ſame diſtances

 from

from B, C, and A, muſt be always at the ſame diſtances from O, that is, OR, OS, and OQ, are conſtant, and the point O at reſt. And this muſt alſo be the caſe if *a*, *b*, and *c* be variable, provided they have the ſame conſtant ratio amongſt themſelves.

Corollary. Hence the points Q, R, and S, are the nodes of the great circles CQA, ARB, and BSC.

Scholium. The demonſtration of this propoſition being thus ſtrictly given, ſome notion may be obtained of the manner in which the point O varies its place upon the ſpherical ſurface when the velocities along the circles AB, BC, and CA, are variable. Thus, let ſuch ſpherical ſurface, ſo revolving, receive an inſtantaneous impulſe, at the diſtance of a quadrant or 90° from S, in a direction perpendicular to the plane of the great circle CSB; then, the centre of the ſphere may be kept at reſt by an equal and contrary impulſe at this centre; and ſince, by hypotheſis, the impulſe is given 90° from S, and in a direction perpendicular to the plane of the great circle CSB, it can neither alter the place of the node S upon the circle, nor the velocity in the direction of its periphery, but only thoſe in AB and CA. Thus, if the velocity in BA which before was $= c$ be now equal to z; then, as ſ. SB : z :: ſ. SC : the velocity along CA, let this $= y$, whilſt ſtill the velocity along CB continues as before $= a$; and this will cauſe the point O to fall upon another point of the great circle SA: ſo that whereas before the ſines of OS, OR, and OQ, were as a, c, and b, they ſhall now be as a, z, and y. Conſequently, as ſ. SO : rad. $= 1$ :: a : the velocity at 90° from O, ſ. OR : 1 :: z : the velocity at 90° from O, and ſ. OQ : y :: 1 : velocity at 90° from O, which three quantities muſt therefore be equal to one another, and to the angular velocity of the ſphere about the axis whoſe pole is is O; let this angular velocity $=e$, then muft $e\times$ f. SO $=a$, $e\times$ f. OR $=z$, and $e\times$ f. OQ $=y$, and the fum of the fquares of thefe three, or $a^2+z^2+y^2=e^2\times$ f. $SO^2+e^2\times$ f. $RO^2+e^2\times$ f. $QO^2=e^2$, becaufe f. SO^2+ f. RO^2+ f. $QO^2=1$, hence $e=\sqrt{a^2+z^2+y^2}$; whereas, before the impulfe $e=\sqrt{a^2+b^2+c^2}$. Thus not only the place of O, but, if z^2+y^2 be not $=b^2+c^2$, the angular velocity of the fphere about its fingle axis will alfo be altered. Hence then if, inftead of an inftantaneous impulfe, a motive force be fuppofed to act in the fame direction, and meafured at the fame point where the impulfe was juft now fuppofed to act; fuch force can neither vary the point S nor the velocity a, but will in time vary b and c, and caufe the point O to alter its place in SA; and thus the velocities b and c will vary to y and z, and $e=\sqrt{a^2+b^2+c^2}$ to $e=\sqrt{a^2+y^2+z^2}$, juft as it would have been by a fingle impulfe, excepting that then, when the impulfe was over, y and z muft have become conftant quantities, whereas now they will vary perpetually during the time that the motive force acts, and the point O will fhift its place fo as at different times to coincide with different points of AS, though at any one inftant the point of the furface that coincides with it muft be at reft, by Prop. 2.

PROPOSITION IV.

If a fpherical furface, whofe center is at reft, revolve in any manner whatfoever, fo that the velocities along the three quadrants bounding any octant thereof be expreffed by any three variable quantities x, y, and z; to find the neceffarily correfponding accelerating forces with which the place of the natural

natural or momentary axis, and the angular velocity of the ſurface round it are varied.

Here, other things remaining as in the preceding propoſition, inſtead of the conſtant quantities a, b, and c, we have the variable ones x, y, and z. Let the variable ſines and coſines of AO, BO, and CO, be reſpectively expreſſed by b and β, g and γ, and d and δ; and let t = the time from the commencement of the motion; then it is well known, that the reſpective accelerating forces along CB, CA, and AB, muſt be expreſſed by $\frac{\dot{x}}{\dot{t}}$, $\frac{\dot{y}}{\dot{t}}$, and $\frac{\dot{z}}{\dot{t}}$; and the radius of the ſphere being ſuppoſed = unity, the angular velocity about the axis whoſe pole is $O = e = \sqrt{x^2+y^2+z^2} = e\sqrt{\beta^2+\gamma^2+\delta^2}$, $e\beta = x$, $e\gamma = y$, $e\delta = z$, $\dot{x} = e\dot{\beta} + \beta\dot{e}$, $\dot{y} = e\dot{\gamma} + \gamma\dot{e}$, $\dot{z} = e\dot{\delta} + \delta\dot{e}$, $\beta^2+\gamma^2+\delta^2 = 1$, $\beta\dot{\beta} + \gamma\dot{\gamma} + \delta\dot{\delta} = 0$, $\beta^2+\gamma^2 = 1-\delta^2 = d^2$, $\beta^2+\delta^2 = 1-\gamma^2 = g^2$, $\gamma^2+\delta^2 = 1-\beta^2 = b^2$, $\dot{e} = \frac{x\dot{x}+y\dot{y}+z\dot{z}}{\sqrt{x^2+y^2+z^2}} = \beta\dot{x}+\gamma\dot{y}+\delta\dot{z} = \frac{\dot{x}-e\dot{\beta}}{\beta} = \frac{\dot{y}-e\dot{\gamma}}{\gamma} = \frac{\dot{z}-e\dot{\delta}}{\delta}$. And, by ſpherics, as $g : 1 :: \delta : \frac{\delta}{g}$ = ſ. OBR = ſ. QA $:: \beta : \frac{\beta}{g}$ = ſ. OBS = ſ. CQ = coſ. AQ, tang. AQ = $\frac{\delta}{\beta}$ and the fluxion of the arc AQ = $\frac{\beta\dot{\delta}-\delta\dot{\beta}}{\beta^2+\delta^2}$. But, by the foregoing propoſition, BO is perpendicular to Bb the track of the point B; conſequently, as ſ. OBR = ſ. AQ : ſ. OBS = coſ. AQ $:: z : x$; therefore the tangent of AQ = $\frac{z}{x}$ and $\dot{\text{AQ}} = \frac{x\dot{z}-z\dot{x}}{x^2+z^2} = \frac{e\beta \times \overline{e\dot{\delta}+\delta\dot{e}} - e\delta \times \overline{e\dot{\beta}+\beta\dot{e}}}{e^2\beta^2+e^2\delta^2} = \frac{\beta\dot{\delta}-\delta\dot{\beta}}{\beta^2+\delta^2}$ as before; therefore, whether e be conſtant or variable it makes no difference in the expreſſion for $\dot{\text{AQ}}$. In like manner it will appear, that $\dot{\text{BR}} = \frac{y\dot{x}-x\dot{y}}{x^2+y^2} = \frac{\gamma\dot{\beta}-\beta\dot{\gamma}}{\beta^2+\gamma^2}$, and $\dot{\text{CS}} = \frac{z\dot{y}-y\dot{z}}{y^2+z^2} = \frac{\delta\dot{\gamma}-\gamma\dot{\delta}}{\gamma^2+\delta^2}$. Moreover, as rad. = 1 : the alteration of the place of Q round B, or in the great circle $AC = A\dot{Q}$ :: f. $BO = g = \sqrt{\beta^2 + \delta^2}$: $\frac{\beta\dot{\delta} - \delta\dot{\beta}}{\sqrt{\beta^2 + \delta^2}}$ = the momentary alteration of place of O round B, or in a direction perpendicular to the great circle BOQ at O, and the corresponding alteration of BO, that is, $B\dot{O} = -\frac{\dot{\gamma}}{\sqrt{\beta^2 + \delta^2}} = -\frac{\dot{\gamma}}{g}$, the fluxion therefore of the track of O upon the spherical surface $= \sqrt{\frac{\beta^2\dot{\delta}^2 - 2\beta\dot{\beta}\delta\dot{\delta} + \delta^2\dot{\beta}^2 + \dot{\gamma}^2}{g^2}} = \sqrt{\frac{\beta^2\dot{\delta}^2 - 2\beta\dot{\beta}\delta\dot{\delta} + \delta^2\dot{\beta}^2 + \dot{\gamma}^2 \times \overline{\beta^2 + \gamma^2 + \delta^2}}{\beta^2 + \delta^2}} =$ $\sqrt{\frac{\beta^2\dot{\delta}^2 - 2\beta\dot{\beta}\delta\dot{\delta} + \delta^2\dot{\beta}^2 + \beta^2\dot{\gamma}^2 + \delta^2\dot{\gamma}^2 + \beta^2\dot{\beta}^2 + 2\beta\dot{\beta}\delta\dot{\delta} + \delta^2\dot{\delta}^2}{\beta^2 + \delta^2}} = \sqrt{\dot{\delta}^2 + \dot{\beta}^2 + \dot{\gamma}^2}$. Again, the accelerating force in $BA = \frac{\dot{z}}{\dot{t}}$ resolved into the direction of the great circle BO at B is $\frac{\dot{z}}{\dot{t}} \times$ cof. $OBR = \frac{\dot{z}}{\dot{t}} \times \frac{\beta}{g}$, and that $\frac{\dot{x}}{\dot{t}}$ along BC resolved into the same direction is $\frac{\dot{x}}{\dot{t}} \times \frac{\delta}{g}$, and the difference of these, or the accelerating force in the direction BO in the sense from O towards $B = \frac{\beta\dot{z} - \delta\dot{x}}{g\dot{t}} = \frac{\beta \times \overline{e\dot{\delta} + \delta\dot{e}} - \delta \times \overline{e\dot{\beta} + \beta\dot{e}}}{g\dot{t}} =$ $e \times \frac{\beta\dot{\delta} - \delta\dot{\beta}}{g\dot{t}}$; in like manner that along CO in the sense from O towards $C = \frac{\gamma\dot{x} - \beta\dot{y}}{d\dot{t}} = e \times \frac{\gamma\dot{\beta} - \beta\dot{\gamma}}{d\dot{t}}$, and that along OA from O towards $A = \frac{\delta\dot{y} - \gamma\dot{z}}{b\dot{t}} = e \times \frac{\delta\dot{\gamma} - \gamma\dot{\delta}}{b\dot{t}}$; and as f.ROA (fig. 3.) = f.COA $= \frac{\gamma}{bd}$: 1 :: this last mentioned force : $de \times \frac{\delta\dot{\gamma} - \gamma\dot{\delta}}{\gamma\dot{t}}$ = one equivalent thereto, but acting perpendicular to CO, and urging from O, that is, drawing the great circle DOE perpendicular to BO; then, as 1 : f. DOC = f. ROE $= \frac{\delta\gamma}{gd}$ :: this last force : the

same

ſame reduced into the direction OE $= e\delta \times \frac{\delta\dot{\gamma}-\gamma\dot{\delta}}{g\dot{t}}$ acting perpendicular to the great circle BO, and in the ſenſe from O towards E : the ſame force reduced into the direction of the great circle BO at O is $= e\beta \times \frac{\delta\dot{\gamma}-\gamma\dot{\delta}}{g\gamma\dot{t}}$ in the ſenſe from O towards Q : in like manner is found a force equivalent to that in CO, but acting perpendicular to AO $= ebd \times \frac{\gamma\dot{\beta}-\beta\dot{\gamma}}{d\gamma\dot{t}}$, which reduced into the direction OD is $= e\beta \times \frac{\gamma\dot{\beta}-\beta\dot{\gamma}}{g\dot{t}}$ in the ſenſe from O towards D; but this ſame force perpendicular to AO, when reduced into the direction BO, is $= e\delta \times \frac{\gamma\dot{\beta}-\beta\dot{\gamma}}{g\gamma\dot{t}}$ in the ſenſe from O towards Q, which being added to the other above found force in BO gives $\frac{e\delta \times \overline{\gamma\dot{\beta}-\beta\dot{\gamma}} + e\beta \times \overline{\delta\dot{\gamma}-\gamma\dot{\delta}}}{g\gamma\dot{t}} = -e \times \frac{\beta\dot{\delta}-\delta\dot{\beta}}{g\dot{t}} =$ the accelerating force ariſing from thoſe which act at O along the great circles OA, OC, which force acts in the ſenſe from O towards Q, and therefore in a contrary ſenſe, that is, from O towards B it muſt be $= e \times \frac{\beta\dot{\delta}-\delta\dot{\beta}}{g\dot{t}}$ as before found, the operation thus proving itſelf. In like manner, from the two forces now found, which act perpendicular to OB at O, there muſt ariſe one acting along OD in the ſenſe from O towards D, which will therefore be $=$ $\frac{e\beta \times \overline{\gamma\dot{\beta}-\beta\dot{\gamma}} - e\delta \times \overline{\delta\dot{\gamma}-\gamma\dot{\delta}}}{g\dot{t}} = \frac{e}{g\dot{t}} \times \overline{\gamma\beta\dot{\beta}-\beta\beta\dot{\gamma}-\delta\delta\dot{\gamma}+\gamma\delta\dot{\delta}} = \frac{e}{g\dot{t}} \times \overline{-\gamma\dot{\gamma}^{2}-\beta^{2}\dot{\gamma}-\delta^{2}\dot{\gamma}} = -\frac{e\dot{\gamma}}{g\dot{t}}$. This laſt force may be otherwiſe found thus, the acceleration $= \dot{y}$ round B at Q, and as $1 : g :: \dot{y} : g\dot{y} =$ the acceleration round B at O owing to $\dot{y}$, in like manner, the acceleration round C at O owing to $\dot{z}$ is $= d\dot{z}$, which reſolved

 into

into the direction perpendicular to BO at O is $= d\dot{z} \times$ ſ. ROE $= \frac{\gamma\delta\dot{z}}{g}$, alſo the acceleration $b\dot{x}$ at O perpendicular to AO reduced into the direction perpendicular to BO $= b\dot{x} \times$ ſ. DOS $= \frac{\gamma\beta\dot{x}}{g}$, hence the whole acceleration along DE at O, which manifeſtly ariſes from theſe three, is $= \frac{\gamma\beta\dot{x}}{g} + \frac{\gamma\delta\dot{z}}{g} - g\dot{y}$, and the accelerative force $= \frac{\gamma\beta\dot{x} + \gamma\delta\dot{z} - g^2\dot{y}}{gi}$ which, properly reduced, becomes $- \frac{e\dot{\gamma}}{gi}$ as before. And the force which is compounded of the two forces $e \times \frac{\beta\dot{\delta} - \delta\dot{\beta}}{gi}$ and $- \frac{e\dot{\gamma}}{gi}$ is $= \frac{e}{gi}\sqrt{\overline{\beta\dot{\delta} - \delta\dot{\beta}}^2 + \dot{\gamma}^2} = \frac{\dot{e}}{i}\sqrt{\dot{\beta}^2 + \dot{\gamma}^2 + \dot{\delta}^2}$ acting perpendicular to the track of O upon the moving ſpherical ſurface; and $\frac{\dot{e}}{i} = \frac{\beta\dot{x} + \gamma\dot{y} + \delta\dot{z}}{i}$ is the accelerating force acting along the midcircle, or that which is 90° diſtant from O, to alter the velocity about the natural or momentary axis whoſe pole is O. Hence, anſwerable to the three accelerating forces $\frac{\dot{x}}{i}$, $\frac{\dot{y}}{i}$, and $\frac{\dot{z}}{i}$, round the axes whoſe poles or ends A, B, and C, are always the ſame invariable points upon the moving ſpherical ſurface, there ariſe three other accelerating forces, namely, $e \times \frac{\beta\dot{\delta} - \delta\dot{\beta}}{gi}$, $- \frac{e\dot{\gamma}}{gi}$, and $\frac{\beta\dot{x} + \gamma\dot{y} + \delta\dot{z}}{i}$; the two former acting at the pole of the momentary axis, and the latter is that whereby the velocity about the momentary axis is altered.

SCHOLIUM I.

From the preceding inveſtigation of the forces $e \times \frac{\beta\dot{\delta} - \delta\dot{\beta}}{gi}$ and $- \frac{e\dot{\gamma}}{gi}$, it follows, that they are not at all affected in expreſſion by the

the variation of e, but are denoted by the same quantities, whether e be conſtant or variable; which concluſion, and alſo the values of the forces themſelves, is perfectly agreeable to what is brought out by Mr LANDEN, by a method ſo very different, in the Philoſophical Tranſactions for 1777.

But it is here carefully to be noted, that theſe are not motive forces, but accelerative ones; for no notice whatever is yet taken of the internal ſtructure of the revolving globe; but the expreſſions hold true, be that ſtructure what it will: if it be ſuch that one and the ſame quantity, drawn into each accelerating force, will give the correſpondent motive one, then are the motive forces proportional to the accelerative ones, but otherwiſe not. It may here alſo be obſerved, that it is quite conformable to nature, that theſe accelerating forces ſhould be expreſſed by the ſame quantities whether e be conſtant or variable; for theſe forces, acting at the pole of the natural axis, cannot poſſibly have any effect upon the velocity round it. But it is not hence by any means to be concluded, that the velocity about the axis is therefore conſtant; becauſe theſe are not, in general, the only accelerating forces that act upon the body, but there is alſo a third accelerating force whoſe value is $\frac{\dot{e}}{t}$ ariſing from the different variability of x, y, and z, and which cannot vaniſh except $\beta\dot{x}+\gamma\dot{y}+\delta\dot{z}=0$, it therefore can only vaniſh in particular caſes.

If the equation $\dot{e}=\beta\dot{x}+\gamma\dot{y}+\delta\dot{z}$ be ſquared, there will thence ariſe after due ordering $\dot{e}^2=\dot{x}^2+\dot{y}^2+\dot{z}^2-e^2\times(\overline{\gamma\dot{\beta}-\beta\dot{\gamma}}|^2+\overline{\beta\dot{\delta}-\delta\dot{\beta}}|^2+\overline{\delta\dot{\gamma}-\gamma\dot{\delta}}|^2)$, where the member which is drawn into e^2 keeps its form whether e be conſtant or variable, but by no means will $\dot{x}^2+\dot{y}^2+\dot{z}^2$, after due ſubſtitution, do ſo

 too.

too. If $e=0$, then $\dot{e}^2 = \dot{x}^2 + \dot{y}^2 + \dot{z}^2$, the motion being then round what M. EULER and M. D'ALEMBERT call the *initial axis*, or that about which the body at reſt would be firſt urged to move by any external forces acting upon it; and which they have determined with ſo much labour; though here it follows, as a neceſſary conſequence, that the force with which the body is turned round this initial axis is $= \sqrt{\frac{\dot{x}^2}{\dot{t}^2} + \frac{\dot{y}^2}{\dot{t}^2} + \frac{\dot{z}^2}{\dot{t}^2}}$, or a force = the ſum of the forces round the axes whoſe poles are A, B, and C.

Moreover, by the general laws of all motion, $\frac{\beta\dot{\delta} - \delta\dot{\beta}}{g\dot{t}}$, $-\frac{\dot{\gamma}}{g\dot{t}}$, and $\sqrt{\frac{\dot{\beta}^2 + \dot{\gamma}^2 + \dot{\delta}^2}{\dot{t}^2}}$ are the velocities with which the pole of the momentary axis ſhifts its place in directions perpendicular to BO, along BO and along its own track on the ſurface reſpectively. And it is by taking the fluxions of theſe, and dividing each fluxion by that of the time, that the accelerating forces are had, which are due to ſuch alteration of place of the momentary pole; and theſe muſt by no means be confounded with the forces before found $-\frac{e\dot{\gamma}}{g\dot{t}}$ and $\frac{e}{g\dot{t}} \times \overline{\beta\dot{\delta} - \delta\dot{\beta}}$ in thoſe directions, theſe laſt pertaining to the tendency of the ſurface itſelf to motion at O, and the others to the ſhifting of the pole of the axis upon the ſurface, which are different motions, as will more clearly appear from what follows.

The preceding general properties of motion obtain in all bodies revolving round a center at reſt, be their motion ever ſo irregular; the three great circles bounding an octant of the ſpherical ſurface revolving with the body are alſo taken *ad libitum*, being any ſuch circles whatever upon the ſurface; and hence the following very important conſequence is drawn, *viz.*

If any body be in motion, or put in motion, by inſtantaneous impulſe or otherwiſe, about its center of gravity at reſt in abſolute ſpace, if, by any means, the accelerating forces acting along the three great circles bounding any octant of a ſpherical ſurface that has the ſame center of gravity and revolves with the body, can be found, thoſe acting at every other point of ſuch ſurface will neceſſarily follow as natural conſequences of theſe three, and thus all the motions of ſuch body will be abſolutely determined.

SCHOLIUM II.

As the above concluſions are exceedingly general, in order to form a diſtinct idea how ſuch ſurface moves, it may be proper here to illuſtrate it by a particular example. Let then the velocity x be ſuppoſed conſtant, and alſo the angular velocity e; then, from what is ſhewn above, ſince $x\dot{x} = 0$, $e^2 = \dot{x}^2 + y^2 + z^2 = e^2 \times \overline{\beta^2 + \gamma^2 + \delta^2}$, $e\dot{e} = y\dot{y} + z\dot{z} = 0 = e^2 \times \overline{\gamma\dot{\gamma} + \delta\dot{\delta}}$, $\gamma\dot{\gamma} + \delta\dot{\delta} = 0$, $\dot{\beta} = 0$, β a conſtant quantity, therefore b is conſtant, and the track of the point O upon the ſurface is a leſſer circle of the ſphere at the conſtant diſtance AO from the invariable point A of the ſurface, the radius of ſuch leſſer circle being $= b =$ ſ. AO (fig. 4.), alſo $y^2 + z^2 =$ the conſtant quantity $e^2 - x^2 = e^2 - e^2\beta^2 = e^2 b^2 = e^2 \times \overline{\gamma^2 + \delta^2}$, $z\dot{z} = -y\dot{y}$, $\delta\dot{\delta} = -\gamma\dot{\gamma} = g\dot{g}$, and the velocity $\sqrt{\frac{\dot{\beta}^2 + \dot{\gamma}^2 + \dot{\delta}^2}{\dot{t}^2}}$ with which the pole O ſhifts its place $= \sqrt{\frac{\dot{\gamma}^2 + \dot{\delta}^2}{\dot{t}^2}} = \sqrt{\frac{\delta^2\dot{\delta}^2 + \gamma^2\dot{\delta}^2}{\gamma^2\dot{t}^2}} = \frac{b\dot{\delta}}{\gamma\dot{t}} = \frac{b\dot{\delta}}{\dot{t}\sqrt{b^2 - \delta^2}} = \frac{\dot{\text{EO}}}{\dot{t}}$. But ſtill an expreſſion for $\dot{t}$ is wanting; to the two preceding *data* it is therefore neceſſary to add a third, which may be that the velocity with which O ſhifts its place in the circle EOF is alſo conſtant. Which will

will come to the ſame as a caſe occurring hereafter, when $\frac{\dot{x}}{\dot{t}}=0$, $\frac{\dot{y}}{\dot{t}}=-\frac{xz}{A}$ and $\frac{\dot{z}}{\dot{t}}=\frac{xy}{A}$ where A is ſome conſtant quantity; for then $\dot{x}=0=e\dot{\beta}+\beta\dot{e}$ $e\dot{e}=x\dot{x}+y\dot{y}+z\dot{z}=y\dot{y}+z\dot{z}=e\gamma\dot{y}+e\delta\dot{z}$, $\dot{e}=\gamma\dot{y}+\delta\dot{z}=-\frac{\gamma xz\dot{t}}{A}+\frac{\delta xy\dot{t}}{A}=\frac{e^2\dot{t}}{A}\times\overline{-\gamma\beta\delta+\gamma\beta\delta}=0$, therefore e is conſtant, and $\dot{t}=\frac{A\dot{z}}{xy}=\frac{Ae\dot{\delta}}{e^2\beta\gamma}=\frac{A\dot{\delta}}{e\beta\gamma}$, and ſince $\dot{e}=0$, and $\dot{x}=e\dot{\beta}+\beta\dot{e}=0=e\dot{\beta}$; therefore $\dot{\beta}=0$, β conſtant, and $\gamma=\sqrt{b^2-\delta^2}$; therefore $\dot{t}=\frac{A}{eb\beta}\times\frac{b\dot{\delta}}{\sqrt{b^2-\delta^2}}$, and $t=\frac{A}{eb\beta}\times$ arc EO; conſequently, the velocity with which O ſhifts its place in the arch EO is $=\frac{eb\beta}{A}$, which is a conſtant quantity.

PROPOSITION V.

The ſame being given, as in the laſt propoſition, it is propoſed to illuſtrate the manner in which the ſurface moves with reſpect to a point at reſt in abſolute ſpace.

Let Z (fig. 4.) be a point touching the ſurface, but at reſt in abſolute ſpace whilſt the ſurface moves under it in any manner whatſoever. In any one poſition of the octant ABC through Z draw the great circles A*s*, B*q*, and C*r*, which by the property of the ſphere muſt be perpendicular to BC, CA, and AB, reſpectively; then muſt the velocities of the ſpherical ſurface at *s*, *q*, and *r*, in directions perpendicular to each of the circles A*s*, B*q*, and C*r*, be *x*, *y*, and *z*, the angular velocities therefore about Z, with which the ſurface paſſes under *s*, *q*, and *r*, muſt be $\frac{x}{\text{ſ. }Zs}$, $\frac{y}{\text{ſ. }Zq}$, and $\frac{z}{\text{ſ. }Zr}$; through Z and O draw the quadrant of a great circle ZY; then, as $\beta : x ::$ ſ. OY

: $e \times$ ſ. OY = the velocity of the moving ſpherical ſurface at Y, which is therefore the angular velocity of the ſurface at Y round an axis at reſt whoſe pole is Z, becauſe ZY = 90°; which four values obtain, let the point Z be taken at reſt in abſolute ſpace whereſoever it will. Alſo, $e \times$ ſ. OZ is the velocity with which the ſurface paſſes under Z in a direction perpendicular to the great circle OZ at Z, which muſt therefore be the real velocity of the ſurface itſelf there at that inſtant; therefore the fluxion of the track upon the ſurface which continually paſſes under Z is $= e \times$ ſ. OZ $\times \dot{t} =$ $\sqrt{\text{ſ. Z}\dot{s}^2 + \text{ſ. Z}\dot{q}^2 + \text{ſ. Z}\dot{r}^2}$ From which equation, and the properties of O found in the preceding propoſitions, general expreſſions for the relation of Z and O may be obtained. But, ſeeing that there is ſuch a latitude in determining or fixing upon a proper point Z out of an infinity of points at reſt, and this handled in a general manner will run into a complex *calculus*; in order to fix upon a point Z under the moſt eligible conditions, it may be beſt to deduce them from the properties of any particular problem that comes under conſideration.

For example, taking that in the ſecond ſcholium to the laſt propoſition, where x and e are conſtant, and $y^2 + z^2 = e^2 - x^2$ is alſo conſtant and $= e^2\gamma^2 + e^2\delta^2 = e^2 - e^2\beta^2 = e^2b^2$, or $\gamma^2 + \delta^2 = b^2$ alſo conſtant; and the velocity with which O ſhifts its place along its proper track $= \frac{eb\beta}{A}$, conſtant alſo. Here, in order to fix upon a proper point Z, ſuppoſe the motion to begin when O (fig. 5.) is upon the great circle AB at E, and after ſome determinate time $= t$, ſuppoſe the octant ABC to have arrived in the poſition A′B′C′, and that in this time the point O has ſhifted its place from E to O, that is, ſuppoſing the octant ABC to be at reſt in abſolute ſpace, while A′B′C′ is in motion, on A′B′

 taking

taking $A'e = AE$, the point O will have ſhifted its place in the time t, in abſolute ſpace from E to O; and upon the moving ſpherical ſurface from e to O along a leſs circle whoſe radius is equal to the ſine of $AE = $ ſ. $A'e = $ ſ. $A'O = b$. Now, at the commencement of the motion, that is, at AB, the firſt velocity of the point A along CA is $e \times$ ſ. $AE = eb =$ the then value of y, becauſe the pole of the natural axis of motion E being then upon AB, the value of $z = 0$, and the pole E ſhifting its place in the ſenſe EO in abſolute ſpace, and the invariable point A of the ſpherical ſurface moving in the ſenſe AA′, there muſt be ſome point Z between E and A at reſt with reſpect to both theſe motions, or round which both of them may be ſuppoſed performed; its property muſt then be ſuch, that as ſ. AZ : ſ. EZ :: velocity of $A = eb$: to the velocity with which the pole E begins to ſhift its place $= eb \times \frac{\text{ſ. EZ}}{\text{ſ. AZ}}$; but $\frac{eb\beta}{A}$ is the velocity with which it ſhifts its place upon the moving ſpherical ſurface at E about the radius $=$ ſ. AE $=$ ſ. $\overline{AZ + EZ}$. And when O is at E the velocity with which the ſpherical ſurface paſſes under Z will be $e \times$ ſ. EZ. Again, when O is the place of the momentary pole, the velocity of the point $A' = \sqrt{y^2 + z^2} = e\sqrt{\gamma^2 + \delta^2} = eb$ as before, and the velocity with which O ſhifts it place round $A' = \frac{eb\beta}{A}$ as before; it therefore ſhifts its place round ſome point Z in abſolute ſpace ſuch that $eb \times \frac{\text{ſ. OZ}}{\text{ſ. AZ}}$ is ſtill the velocity with which it ſhifts it, which, becauſe $A'O = AE$, muſt be the ſame velocity and the ſame point Z as before. Conſequently, the point O ſhifts its place along a leſſer circle of the ſphere whoſe radius $=$ ſ. EZ $=$ ſ. OZ, and in the time of ſuch ſhifting from E to O, or from A to A′, the point of the ſurface which at firſt was under Z will arrive at z in A′B′, where $A'z = AZ$, and E conſidered as the ſame invariable point of the ſurface will arrive at e, ſo that $A'e = AE$; therefore, ſince $EZ = OZ$ is conſtant, and Z at reſt both with with reſpect to the velocity eb of A′ round it, and the velocity $\frac{eb\beta}{A}$ with which O ſhifts its place, it muſt be as ſ. EZ = ſ. OZ : ſ. AZ :: $\frac{eb\beta}{A} : eb :: \frac{\beta}{A} : 1$, but b and β are the ſine and coſine of $AE = A'O = EZ + AZ$; therefore, as ſ. EZ = $b \times$ coſ. AZ $- \beta \times$ ſ. AZ : $\beta \times$ ſ. AZ :: $\frac{1}{A} : 1$, and as $b \times$ coſ. AZ : $\beta \times$ ſ. AZ :: $\frac{1}{A} + 1 : 1 :: \frac{b}{\beta} =$ tang. AE : tang. AZ $= \frac{b}{\beta} \times \frac{A}{A+1}$, and ſ. AZ : coſ. AZ = ſ. BZ :: $bA : \overline{A+1} \times \beta$ :: tang. AE : $\frac{A+1}{A}$, ſ. AZ $=$

$$\frac{Ab}{\sqrt{A^2 + 2A\beta^2 + \beta^2}}, \text{ and coſ. AZ} = \frac{\overline{A+1} \times \beta}{\sqrt{A^2 + 2A\beta^2 + \beta^2}};$$

and thus a diſtinct idea of this motion of the ſpherical ſurface is obtained, it being now clear, that the point A′ moves round Z at reſt with the velocity eb, and as ſ. ZA : 1 :: $eb : \frac{e\sqrt{A^2 + 2A\beta^2 + \beta^2}}{A} =$ the angular velocity with which A′ moves round the axis whoſe pole is Z, which is therefore conſtant; and at the ſame time the ſurface itſelf moves in the direction of the great circle B′C′, that is about the axis whoſe pole is A′ with the conſtant velocity $x = e\beta$, which two motions may be conſidered as ſeparate, and the reſt as conſequences of them; that is, the point Z is at reſt, and the point A′ moves uniformly round it, whilſt the ſurface upon which A′ is an invariable point moves round the axis whoſe pole is A′ with an uniform angular velocity, theſe two angular velocities being in the ratio of

$\frac{\sqrt{A^2+2A\beta^2+\beta^2}}{A} : \beta$, or of $\sqrt{A^2b^2+\overline{A+1}|^2\times\beta^2} : A\beta$; therefore, the times being inversely as the velocities, as $A\beta : \sqrt{A^2+2A\beta^2+\beta^2}$ :: the time of one revolution of A' round Z : the time of one revolution of the surface round A', that is, round the axis whose pole is A', which time is given because $x=e\beta$, and consequently the time of one revolution of A' round Z is given. Again, $e\times$ s. $OZ = \frac{eb\beta}{\sqrt{A^2+2A\beta^2+\beta^2}}$ = the velocity with which the surface passes under Z (at rest). The angular velocity round the axis whose pole is $O=e$, and the velocity round O in a circle whose radius is $b=be$, O shifts its place in a circle of the same radius b with a velocity $=\frac{eb\beta}{A}$; the time therefore in which O shifts through the lesser circle eO is to that of one revolution round O (which time may be supposed given) $=T$ as $eb :: \frac{eb\beta}{A}$, or as $1 : \frac{\beta}{A}$, that is, as $\frac{\beta}{A} : 1 :: T : \frac{AT}{\beta}$ = the time in which O makes one revolution upon the surface. And as $e\beta : T :: e : \frac{T}{\beta}$ = the time in which the surface makes one revolution round A', or the axis whose pole is A'; and from the analogy above, the time of one revolution of A' round $Z = \frac{AT}{\sqrt{A^2+2A\beta^2+\beta^2}}$. Also, as $1 :$ s. $AZ ::$ $e\beta : \frac{Aeb\beta}{\sqrt{A^2+2A\beta^2+\beta^2}}$ = the velocity with which the surface would pass under Z, owing to the motion only round the axis whose pole is A', and in a sense from B' towards C'; whereas, owing to the compound motion it really moves under Z in a contrary sense with the velocity $\frac{eb\beta}{\sqrt{A^2+2A\beta^2+\beta^2}}$; this is, however, only a necessary consequence of the centres of the circles whose radii are ſ. A′Z and ſ. EZ ſhifting their places in abſolute ſpace, which therefore can in no wiſe affect the velocities round thoſe centres, which velocities muſt ſtill be the ſame relatively to the centres as if the centres were at reſt. Hence, then, the nature of this ſpherical motion is ſuch, that the axis whoſe pole is Z being abſolutely at reſt, the pole O ſo ſhifts its place in a circle whoſe radius = ſ. ZO alſo at reſt, as to do ſo with a conſtant velocity $= eb \times \frac{\text{ſ. EZ}}{\text{ſ. AZ}} = \frac{eb\beta}{A} =$ the velocity with which it ſhifts its place in the circle *e*O on the moving ſurface, the track therefore on the moving ſurface oſculates or rolls upon that on the immoveable one. Therefore, ſince $\frac{AT}{\beta} =$ the time of one revolution of O upon the moving ſurface, and the time of one revolution of A′, and conſequently O round Z $=$ $\frac{AT}{\sqrt{A^2+2A\beta^2+\beta^2}}$; in the time of one revolution of O on the moving ſurface, it will have ſhifted its place round Z in the circle whoſe radius = ſ. OZ, through an arc = the whole periphery $\times \frac{\sqrt{A^2+2A\beta^2+\beta^2}}{\beta}$, that is, it will have made $\sqrt{\frac{A^2}{\beta^2}+2A+1}$ revolutions round Z: for, as the two circles *e*O and EO oſculate, it will take $\frac{\text{ſ. OA}}{\text{ſ. OZ}} = \sqrt{\frac{A^2}{\beta^2}+2A+1}$ times the periphery of EO to go round *e*O, that is, the point A′, and conſequently O will have moved this number of times round Z at reſt, whilſt O ſhifts its place once round the ſpherical ſurface in motion. Hence then the nature of the motion round the momentary axis whoſe pole is O, and the fixed one whoſe pole is Z, will be apparent from the following ſimple contrivance. A circle EO to radius = ſ. ZO = ſ. ZE being drawn upon a ſpherical ſurface at reſt, an octant of which is ABC, let a paper, or other

other loose surface, be fitted to this octant, and having on the centre A and radius AE described a circular arc on the loose surface, let the part thereof EOFCBE be cut away, and completing the circle EF of the remaining part, let the circumference of this circle be moved uniformly along the circumference of the less fixed circle EO with the celerity $\frac{eb\beta}{A}$ beginning at the point E in each, so that the moving circle may roll along the fixed one, that is, so that the arc O*e* of the moving circle which has been in contact with the fixed one may be always equal to the arc EO of the fixed one with which it has been in contact; then, since OZ and ZA′ are constant, and OZ perpendicular to both circles, the point A′ must describe upon the fixed surface, the same *locus* as in the case of the motion above specified. The *locus* also of the momentary pole O will be the same, and the angular velocity of A′ about the momentary axis the same as that of the moving surface about it: for the celerity of O about the axis whose pole is $Z = \frac{eb\beta}{A}$ being equal to the celerity about A′ in motion, and the *locus* of A′ being a circle whose radius = s. ZA, we have, as s. ZO : s. ZA′ :: $\frac{eb\beta}{A}$: eb = the velocity of A′, and as s. OA′ = b : eb :: radius = 1 : e = the velocity about the momentary axis, as it ought.

From this complete solution of the particular case may be collected in general, that if the octant ABC be taken such upon the moving spherical surface, that the track of O thereupon may cross the two great circles AB and AC at right angles, a point which is at rest with respect to both motions, or round which they are performed like a single motion, may at the instant of the momentary pole's crossing each of those great

great circles be found, in the ſame manner as in the particular caſe here ſpecified. And it will alſo be found for any poſition of O, by means of the expreſſions for the velocities found in Scholium I. Prop. IV.; but of this more hereafter.

PROPOSITION VI.

If a parallelopipedon (or other * ſolid) revolving uniformly with an angular velocity $= z$ about one of its permanent axes of rotation, receive an inſtantaneous impulſe in a direction parallel to that axis, the centre of gravity of the body being ſuppoſed to be kept at reſt by an equal and contrary impulſe given to it, and no other force acting upon the body, it is propoſed to determine the alteration in the motion thereof, in conſequence of ſuch inſtantaneous impulſe.

The impulſe being, by hypotheſis, given in a direction perpendicular to that of the then only motion of every particle of the body, cannot inſtantly alter its angular velocity about the permanent axis; but its immediate effect muſt be to cauſe the body to revolve about a freſh axis, whilſt the angular velocity, and conſequently the *momentum* of rotation about the firſt or permanent axis, remain unaltered by ſuch inſtantaneous impulſe; for though it gives a different direction and velocity to the particles, by cauſing them to revolve about another axis, yet muſt their relative velocity about the firſt remain unaltered by the nature of relative motion, becauſe the ſecond or additional motion is given in a direction perpendicular to the firſt. Any alteration therefore which may be made in the velocity about the firſt axis, by reaſon of the oblique motion of the particles about it, owing to the then revolution about a freſh axis, muſt be a work of time. And to determine ſuch alteration,

* See the *note* (C) at the end of the Paper.

let

let M = the mafs or folidity, and $2d$, $2c$, and $2b$, be the three dimenfions or length, breadth, and thicknefs of fuch parallelopipedon; then it is known that the momentum of *inertia* round the axis on which the dimenfion $2d$ is taken will be $=\frac{1}{3}M \times \overline{c^2+b^2}$, this being no more than the product of a particle of the body into the fquare of its diftance from fuch axis, when integrated through the whole body, as is now too well known to need the repetition here. Let I (fig. 6.) be the centre of gravity or of *inertia* (they being both one) of fuch parallelopipedon, IB the permanent axis on which the dimenfion $2c$ is taken, CI that on which $2b$ is taken, and a perpendicular to the plane BIC (of the paper) at I that on which $2d$ is taken; then on the centre I defcribing the quadrant BSC, whofe radius BI or CI may be fuppofed unity; if the body once revolve about this laft named axis with an angular velocity $=z$ meafured along the great circle BSC, and no external force or impulfe act upon it, it is agreed and well known, that the centrifugal motive force round fuch axis will be $= Mz^2 \times \frac{c^2+b^2}{3}$ and always being equal in contrary directions round the axis can have no power to alter the place thereof; but fuch motion and motive force continuing always the fame, the axis muft be at reft, and the velocity round it uniform for ever. But, if the body whilft fo revolving receive (as by hypothefis) an impulfe in a direction parallel to this axis, that is, perpendicular to the plane of the circle BCI, and an equal and contrary one to keep the centre I ftill at reft, the faid impulfe being perpendicular to the motion cannot inftantly alter the angular velocity z, but will give the axis itfelf a motion in a plane perpendicular to BCI, and confequently about fome axis SI in the plane BCI, round which axis SI the centrifugal motive forces

forces of the particles being no longer in *equilibrio*, becaufe it is not a permanent axis (except in particular cafes) this oblique motion of the particles will in time alter the velocity z. To determine then the value of the motive force caufing fuch alteration of z, let $ML = 2d$ be a line parallel to the fide of the parallelogram which is a fection of the folid perpendicular to the axis CI, q the middle point of ML, p any other point therein, pm and qn two perpendiculars to the plane which is perpendicular to BCI and paffes through SI; and from B let fall BN perpendicular to the axis SI: then, the point n muft neceffarily fall upon SI, becaufe the plane BSCI produced bifects the folid, join pn which is the perpendicular diftance of p from the axis SI; let $v =$ the velocity of the body at B perpendicular to BI and to the plane BCI (which is the fame invariable one in the body, and that wherein the permanent axes BI and CI are fituated); then, as $BN : v :: 1 :$ the angular velocity of the body about the axis $SI = \frac{v}{BN}$, and by the nature of all motion, as $BN : v :: pn : \frac{pn}{BN} \times v =$ the velocity of the point p round n, or of a particle of the body at p in the circle whofe radius is pn, confequently the centrifugal accelerating force, which is always equal to the fquare of the velocity divided by the radius of motion, is there $= \frac{pn}{BN^2} \times v^2$ acting in the direction pn upon the axis SI, which may be refolved into two others, the one parallel to the plane SmI, which can have no effect in a direction perpendicular to that plane, and the other $= \frac{v^2}{BN^2} \times pm = \frac{v^2}{BN^2} \times qn$ perpendicular to that plane, which drawn into a particle p of the body at p gives $p \times qn \times \frac{v^2}{BN^2} =$ the motive

 force

force of that particle to move the plane S*ml* in a direction parallel to BN, or about the permanent axis which is perpendicular to the plane SBI, and which value is the same in whatever point of ML the particle p is situated.

Let G*g*R (fig. 7.) be a section of the solid by the plane IBSC; then, since the motive force of a particle p of the body situated any where in a line perpendicular to this plane at q is $p \times qn \times \frac{v^2}{BN^2}$, the motive force arising from the dimension ML $=$ $2d$ of the body will be $= 2dv^2 \times \frac{qn}{BN^2}$, and as SI $= 1 :$ In $:: 2dv^2 \times \frac{qn}{BN^2} : 2dv^2 \times \frac{qn \times In}{BN^2}$ = the equivalent motive force acting at the constant distance SI = unity; which must still be integrated with the other two dimensions of the body, because every particle $p = M\dot{p} \times K\dot{R} \times q\dot{g}$. In order to which, let now $f = \frac{v^2}{BN^2}$, s and t = the sine and cosine of QIK = NBI = SC to radius unity, IR $= b$, GK $= c$, KR $= x$, and $qg = y$; then will KI $= x - b$, K$q = y - c$, and as $t :$ KI $:: 1 : q$I $= \frac{x-b}{t} :: s :$ QK $= \frac{s}{t} \times \overline{x-b}$; hence, Q$q$ = Kq − QK $= y - c - \frac{s}{t} \times \overline{x-b}$, and $1 :$ Qq $:: s :$ Q$n = s \times \overline{y-c} - \frac{s^2}{t} \times \overline{x-b} :: t : qn = t \times \overline{y-c} - s \times \overline{x-b}$, and I$n$ = QI + Q$n = s \times \overline{y-c} + t \times \overline{x-b}$; hence $qn \times$ I$n = st \times \overline{y^2 - 2yc + c^2} + t^2 \times \overline{y-c} \times \overline{x-b} - s^2 \times \overline{y-c} \times \overline{x-b} - st \times \overline{x-b}|^2$, which multiplied into $2df\dot{y}$ and the fluent making y only variable so as to comprehend the whole body, when $y = 2c = g$G, is $= 2dfst \times \overline{\frac{2c^3}{3} - 2c \times \overline{x^2 - 2bx + b^2}}$, and this multiplied into $\dot{x}$, and the fluent taken in like manner, will, when $x = 2b$, be $= \frac{8}{3} \times dfst \times$

c[3]

$\overline{c^3b - b^3c} = \text{M}fst \times \frac{c^2 - b^2}{3} = \frac{\text{M}v^2st}{3t^2} \times \overline{c^2 - b^2} = \frac{sv^2}{t} \times \frac{\text{M}}{3} \times \overline{c^2 - b^2}$; but as ſ. BS $= t : v ::$ ſ. CS $= s : \frac{sv}{t} =$ the velocity of the body at C perpendicular to CI and to the plane BCI; let $\frac{sv}{t} = x$ now, and $v = y$, and the preceding fluent becomes $\frac{\text{M}xy}{3} \times \overline{c^2 - b^2}. =$ the motive force acting at S along the circle BSC to alter the velocity z along that circle; and if this be divided by the *inertia* $\frac{\text{M}}{3} \times \overline{c^2 + b^2}$ along BC, it gives $xy \times \frac{c^2 - b^2}{c^2 + b^2} = \frac{\dot{z}}{\dot{t}}$ (where $\dot{t} =$ that of the time) = the accelerating force acting along the circle BC. Now (this being referred to fig. 3.), for the ſame reaſon, as the two velocities x and y along BC and CA turn the body about an axis whoſe pole is R in AB, and thus cauſe the perturbating motive force $\frac{\text{M}xy}{3} \times \overline{c^2 - b^2}$ above computed, muſt the two velocities x and z along BC and BA turn the body about an axis in CA whoſe pole is Q, and proceeding in the very ſame manner as before, the perturbating motive force thence ariſing will be found $= \frac{\text{M}xz}{3} \times \overline{b^2 - d^2}$, to alter the motion along AC, and the accelerative one $= \frac{b^2 - d^2}{b^2 + d^2} \times xz = \frac{\dot{y}}{\dot{t}}$ to alter the velocity y about the permanent axis whoſe pole is B. Alſo, the motive force $\frac{\text{M}yz}{3} \times \overline{d^2 - c^2}$, and the accelerative one $= \frac{d^2 - c^2}{d^2 + c^2} \times yz = \frac{\dot{x}}{\dot{t}}$ to alter the velocity x along BC.

SCHOLIUM I.

Having thus obtained the values of the accelerating forces $\frac{\dot{x}}{\dot{t}}$, $\frac{\dot{y}}{\dot{t}}$, and $\frac{\dot{z}}{\dot{t}}$ (ſee Scholium I. prop. IV.), the matter is now

 brought

brought to an iſſue, and the motions and times may from hence be computed. But it will be proper firſt to ſhew wherein, and why, theſe concluſions differ from thoſe brought out by Mr. LANDEN.

The three perturbating motive forces acting along the peripheries of the three great circles CB, CA, and AB, in fig. 3. Prop. IV. are above found to be $\frac{M}{3}\times\overline{d^2-c^2}\times yz$, $\frac{M}{3}\times\overline{b^2-d^2}\times xz$, and $\frac{M}{3}\times\overline{c^2-b^2}\times xy$ reſpectively, or their equals $\frac{M}{3}\times\overline{d^2-c^2}\times e^2\gamma\delta$, $\frac{M}{3}\times\overline{b^2-d^2}\times e^2\beta\delta$, and $\frac{M}{3}\times\overline{c^2-b^2}\times e^2\beta\gamma$. And if we ſuppoſe the *accelerations* $\dot x$, $\dot y$, and $\dot z$, to be reſpectively proportional to the motive forces, the ſum $\dot x+\dot y+\dot z$ muſt be proportional to the ſum of the three motive forces, and $x\dot x+y\dot y+z\dot z$ or its equal $e\beta\dot x+e\gamma\dot y+e\delta\dot z$ muſt be proportional to $\frac{M}{3}\times\overline{d^2-c^2}\times xe^2\gamma\delta+\frac{M}{3}\times\overline{b^2-d^2}\times ye^2\beta\delta+\frac{M}{3}\times\overline{c^2-b^2}\times ze^2\beta\gamma=\frac{M}{3}\times e^3\beta\gamma\delta\times\overline{d^2-c^2+b^2-d^2+c^2-b^2}$ that is as nothing; conſequently, $e\dot e=x\dot x+y\dot y+z\dot z=0$, in which caſe therefore e muſt be a conſtant quantity. Moreover, theſe quantities now mentioned as reſpectively proportional to one another, turning the equal ratios into equations $\frac{\overline{d^2-c^2}\times yz}{\dot x}=\frac{\overline{b^2-d^2}\times xz}{\dot y}=\frac{\overline{c^2-b^2}\times xy}{\dot z}=\frac{\overline{d^2-c^2}\times e\gamma\delta}{\dot\beta}=\frac{\overline{b^2-d^2}\times e\beta\delta}{\dot\gamma}=\frac{\overline{c^2-b^2}\times e\beta\gamma}{\dot\delta}=\frac{Be\gamma\delta}{\dot\beta}=\frac{De\beta\delta}{\dot\gamma}=\frac{Ce\beta\gamma}{\dot\delta}$; hence $D\beta\dot\beta=-B\gamma\dot\gamma$, and $D\delta\dot\delta=-C\gamma\dot\gamma$, and taking the fluents of theſe two laſt equations, putting n and m for the reſpective values of β and δ when $\gamma=0$, we obtain $D\beta^2=Dn^2-B\gamma^2$, and $D\delta^2=Dm^2-C\gamma^2$, conſequently $\beta=\frac{\sqrt{Dn^2-B\gamma^2}}{D^{\frac{1}{2}}}$ and $\delta=\frac{\sqrt{Dm^2-C\gamma^2}}{D^{\frac{1}{2}}}$; which are the

very

very equations brought out by Mr. LANDEN in ſo very different a manner.

Here then the matter may be ſafely reſted; for the accelerations are moſt certainly as the accelerative forces, and not as the motive ones. Concluſions, therefore, that are drawn from a contrary ſuppoſition cannot be true.

It may not, however, be improper to ſhew here how Mr. LANDEN's motive forces E and E'' ariſe from thoſe above brought out; thus, in fig. 3. Prop. IV. let s and t the ſine and coſine of AQ to radius 1, that is, let $s = \frac{\delta}{g}$ and $t = \frac{\beta}{g}$, then the motive force along BA reſolved into the direction BO becomes $\frac{M}{3} \times \overline{c^2 - b^2} \times e^2 \beta \gamma t$, and that along BC reſolved into the ſame direction BO becomes $\frac{M}{3} \times \overline{d^2 - c^2} \times e^2 \gamma \delta s$, the difference of theſe $= \frac{M}{3} \times e^2 \gamma \times \overline{d^2 - c^2} \times \delta s - \overline{c^2 - b^2} \times \beta t = \frac{Me^2 g\gamma}{3} \times \overline{Ds^2 - C}$ muſt be the motive force acting along the great circle BO in the ſenſe from B towards O, or from O towards Q; and this is the very motive force E determined by Mr. LANDEN, and acting in the ſame manner. The motive force which acts at O perpendicularly to the force E is moſt readily obtained from that acting along CA; for if a tangent be drawn to the great circle BOQ at O (fig. 3.) it will interſect a radius of the ſphere drawn through Q at a diſtance $\left(\frac{1}{g}\right)$ from I the centre of the ſphere = the ſecant of the arc OQ, and as $1 : \frac{1}{g} =$ that ſecant :: the force $\frac{MDe^2\beta\delta}{3}$ acting at the diſtance Q from the centre : $\frac{MDe^2\beta\delta}{3g} = \frac{MDe^2 gst}{3}$ the force acting in the plane of the great circle CIA at the diſtance $\frac{1}{g}$ from the centre I, and perpendicular to a tangent at

O

O to the great circle BOQ; which force being in a direction parallel to and in the ſame plane with the motive force acting at O perpendicular to the ſame tangent muſt be equal to it; that is, the motive force which acts perpendicular to E at O is $= \frac{Me^2g}{3} \times Dst =$ Mr. LANDEN's force E″. And this may alſo be deduced by finding by reſolution the motive forces along CO and AO, and reducing them into the direction of the great circle DOE at O, in the ſame manner as the accelerating forces are managed in Prop. IV. above. Now, theſe forces E and E″ not being the only perturbating ones that diſturb the motion of the body, but others ariſing from the *non-equilibrium* of the particles in motion round the axes which are perpendicular to the planes of the varying momentary great circles BOQ, DOE, they will neither divided by their reſpective *inertia* $\frac{M}{3} \times \overline{b^2 + c^2 + \overline{d^2 - b^2} \cdot s^2}$ and $\frac{M}{3} \times \overline{d^2 + b^2 + \gamma^2 \times \overline{c^2 - b^2 - s^2} \cdot \overline{d^2 - b^2}}$ give the accelerating forces along thoſe circles, nor are proportional to them; but, by the general properties of all motion as proved in Prop. IV. the accelerating forces in thoſe circles are $\frac{\delta}{g} \times \frac{\dot{x}}{\dot{t}} - \frac{\beta}{g} \times \frac{\dot{z}}{\dot{t}}$ ($\dot{t} =$ that of the time) $= \frac{s\dot{x}}{\dot{t}} - \frac{t\dot{z}}{\dot{t}} = \frac{d^2 - c^2}{d^2 + c^2} \times e^2 s\gamma\delta - \frac{c^2 - b^2}{c^2 + b^2} \times e^2 t\beta\gamma = \frac{d^2 - c^2}{d^2 + c^2} \times e^2 g\gamma s^2 - \frac{c^2 - b^2}{c^2 + b^2} \times e^2 g\gamma t^2 = \frac{e\delta\dot{\beta}}{g\dot{t}} - \frac{e\beta\dot{\delta}}{g\dot{t}}$ (S) and $\frac{\gamma\beta}{g} \times \frac{\dot{x}}{\dot{t}} + \frac{\gamma\delta}{g} \times \frac{\dot{z}}{\dot{t}} - \frac{g\dot{y}}{\dot{t}} = \frac{\gamma\beta}{g} \times \frac{d^2 - c^2}{d^2 + c^2} \times e^2\gamma\delta + \frac{\gamma\delta}{g} \times \frac{c^2 - b^2}{c^2 + b^2} \times e^2\beta\gamma + g \times \frac{d^2 - b^2}{d^2 + b^2} \times e^2\beta\delta = \frac{e^2\beta\delta}{g} \times \left(\overline{\frac{d^2 - c^2}{d^2 + c^2} + \frac{c^2 - b^2}{c^2 + b^2} - \frac{a^2 - b^2}{a^2 + b^2}} \times \gamma^2 + \frac{d^2 - b^2}{d^2 + b^2}\right) = \frac{e^2\beta\delta}{g} \times \left(\frac{d^2 - c^2}{d^2 + c^2} \times \frac{c^2 - b^2}{c^2 + b^2} \times \frac{d^2 - b^2}{d^2 + b^2} \times \gamma^2 + \frac{d^2 - b^2}{d^2 + b^2}\right) = -\frac{e\dot{\gamma}}{g\dot{t}}$ (Q). And from theſe equations (S) and (Q) there reſults the analogy, as $\frac{d^2 - c^2}{d^2 + c^2} \times e^2 g\gamma s^2 - \frac{c^2 - b^2}{c^2 + b^2} \times e^2 g\gamma t^2 : \frac{e^2\beta\delta}{g} \times \left(\overline{\frac{d^2 - c^2}{d^2 + c^2} + \frac{c^2 - b^2}{c^2 - b^2} - \frac{d^2 - b^2}{d^2 + b^2}} \times \gamma^2\right.$
+

$+\frac{d^2-b^2}{d^2+b^2}) :: \frac{e\delta\dot{\beta}}{g} - \frac{e\beta\dot{\delta}}{g} : -\frac{e\dot{\gamma}}{g} :: \delta\dot{\beta} - \beta\dot{\delta} : -\dot{\gamma} :: \frac{\delta\dot{\beta}-\beta\dot{\delta}}{g^2} = -\frac{\dot{s}}{t} : -\frac{\dot{\gamma}}{g^2}$; hence, $\frac{d^2-c^2}{d^2+c^2}\times s^2 - \frac{c^2-b^2}{c^2+b^2}\times t^2 : \overline{\frac{a^2-c^2}{d^2+c^2}+\frac{c^2-b^2}{c^2+b^2}}\times\gamma^2+\frac{d^2-b^2}{d^2+b^2}\times g^2 :: -s\dot{s} : \frac{g\dot{g}}{g^2}$, or, putting $\frac{c^2-b^2}{c^2+b^2}=M^2\times\overline{\frac{a^2-c^2}{a^2+c^2}+\frac{c^2-b^2}{c^2+b^2}}$ and $\frac{a^2-c^2}{d^2+c^2}+\frac{c^2-b^2}{c^2+b^2}=N^2\times\overline{\frac{d^2-c^2}{d^2+c^2}+\frac{c^2-b^2}{c^2+b^2}+\frac{b^2-d^2}{d^2+b^2}}$, $-s\dot{s} : \frac{g\dot{g}}{g^2} :: s^2-M^2 : 1-N^2g^2$, or $\frac{-s\dot{s}}{s^2-M^2}=\frac{g\dot{g}}{g^2-N^2g^4}$; but when $g=1$, $s=m$, and the fluents corrected accordingly, give the equation $\frac{1}{2}$ log. of $\frac{m^2-M^2}{s^2-M^2}=\frac{1}{2}$ log. of $\frac{g^2-N^2g^4}{1-N^2}$ − log. of $\frac{1-N^2g^2}{1-N^2}$, consequently, $\sqrt{\frac{m^2-M^2}{s^2-M^2}}=\sqrt{\frac{g^2-N^2g^2}{1-N^2g^2}}$, and $m^2\times\overline{1-N^2g^2}-M^2=s^2g^2\times\overline{1-N^2}-M^2g^2$; but $sg=\delta$, therefore $\overline{m^2-\delta^2}\times\overline{1-N^2}=\overline{M^2-N^2m^2}\times\gamma^2$; or, expunging M and N, $m^2-\delta^2=\frac{\overline{d^2+b^2}\times\overline{c^2-b^2}}{\overline{c^2+b^2}\times\overline{d^2-b^2}}\times\gamma^2-m^2\gamma^2\times\frac{\overline{d^2-c^2}\times\overline{c^2-b^2}}{\overline{d^2+c^2}\times\overline{c^2+b^2}}$ is the equation of the curve which is the locus of O upon the moving ſpherical ſurface; or, if $\frac{a^2+c^2}{d^2-c^2}=A$, $\frac{d^2+b^2}{d^2-b^2}=B$ and, $\frac{c^2+b^2}{c^2-b^2}=C$, $m^2-\delta^2=\frac{B\gamma^2}{C}-\frac{m^2\gamma^2}{AC}$. Which concluſion may be brought out with much more facility, by means of the three original equations above inveſtigated, which expreſs the values of the three accelerating forces $\frac{\ddot{x}}{\dot{t}}$, $\frac{\ddot{y}}{\dot{t}}$, and $\frac{\ddot{z}}{\dot{t}}$, as will be ſhewn hereafter. But it is of importance to have proved here, that this different method when rightly treated comes to the ſame as the other.

SCHOLIUM

SCHOLIUM II.

The velocity of the body in directions of the peripheries of three great circles bounding an octant of the fpherical furface which revolves with it, might have been referred to any other octant befides that whofe angles, as in the preceding folution, are in the poles of the three permanent axes; but then, befides the perturbating force arifing from the motion of the body about each of the three axes whofe poles are in the nodes of the great circles bounding fuch octant, there will, pertaining to each circle, be another perturbating force, arifing from the *non-equilibrium* of the particles of the body in motion, in planes parallel to the plane of each circle, which being confidered would greatly perplex the operation. And hence arifes the neceffity for referring the motion to permanent axes, becaufe about them this laft-mentioned perturbating force vanifhes by reafon of the perfect *equilibrium* of the particles in motion round them; their property being fuch, that if the body begin to move fimply round one of them, it muft uniformly continue fo to do for ever. And if, as in the preceding propofition, the body be compelled to move round fome other axis, ftill during the elementary time $\dot{t}$, notwithftanding that each of thefe axes or their poles has a proper motion of its own, yet the *relative* angular velocity, and confequently the *inertia* and motive force round each axis, will be the fame as if the body revolved with the fingle angular velocity x, y, or z, round only one of them, and confequently fuch velocity can have no power to alter itfelf; but the *equilibrium* of the particles tends to preferve it, for the particles by their motion round one of thefe axes cannot alter the angular velocity about it; but fuch

 alteration

alteration muſt be cauſed by the other motions of the body which are referred to the other two permanent axes as in the foregoing ſolution, and thus produce the forces $\frac{\dot{x}}{\dot{t}}$, $\frac{\dot{y}}{\dot{t}}$, and $\frac{\dot{z}}{\dot{t}}$, acting at the nodes S, Q, and R, of the great circles BC, CA, and AB.

If the two great circles DOE, CQA, be continued, they will meet in a point of the midcircle 90° from O, and make an angle whoſe meaſure is the arc OQ, and if Mr. LANDEN's motive force E″ be reſolved into the direction of the great circle CA, it will become $E'' \times$ coſ. $OQ = E'' \times g = \frac{M}{3} \times \overline{a^2 - b^2} \times e^2 \beta \delta$, the very ſame as inveſtigated in the foregoing propoſition. But Mr. LANDEN's method, beſides the force E″ perpendicular to BO, will likewiſe give two other motive forces perpendicular to AO and CO at O, which reſolved into the directions of the great circles BC and AB will alſo give the above inveſtigated motive forces in thoſe circles, and thus the two methods prove each other.

I know then of no objection but what is already obviated; I ſhall therefore proceed to the ſolution of the following propoſition; firſt, independent of the conſideration of a momentary axis, the properties of which ſhall be inveſtigated afterwards. I could eaſily give the demonſtration that the properties above ſhewn to belong to the parallelopipedon, alſo pertain to any other body; but as this has been done before by Mr. LANDEN, I ſhall take it for granted here.

PROPOSITION VII.

If a body of any form revolve in any manner whatfoever with its centre of gravity at reſt in abſolute ſpace, and ſo as not to be diſturbed by the action of any external force; to determine in what manner it will continue its motion for ever.

Since any body whatever, whoſe permanent axes can be found, may be reduced to an equipollent parallelopipedon which ſhall move in the very ſame manner as the body; let this be ſuppoſed done, M being the maſs or ſolidity of the body, and Ma^2, Mb^2, and Mc^2, the reſpective *momenta* of *inertia* round the three permanent axes of the body whoſe poles in the ſpherical furface whoſe radius is unity revolving as the body revolves and concentric with it are A, B, and C, at the diſtance of a quadrant from each other (fig. 8.); let $x =$ the velocity with which the body moves round the permanent axis whoſe pole is A, and meaſured along the great circle BC at the diſtance of a quadrant from that pole (A) and in the ſenſe from B towards C; in like manner, let $y =$ the velocity round the axis whoſe pole is B, meaſured along CA, and in the ſenſe from C towards A, and $z =$ that round the remaining permanent axis whoſe pole is C meaſured along AB, and in the ſenſe from A towards B. Alſo let $t =$ the time from the commencement of the motion.

Then, the quantities which in the 6th Propoſition were repreſented by $\frac{M}{3} \times \overline{d^2 + c^2}$, $\frac{M}{3} \times \overline{d^2 + b^2}$, $\frac{M}{3} \times \overline{c^2 + b^2}$, $\frac{M}{3} \times \overline{d^2 - c^2}$, $\frac{M}{3} \times \overline{b^2 - a^2}$, $\frac{M}{3} \times \overline{c^2 - b^2}$, $\frac{d^2 - c^2}{a^2 + c^2}$, $\frac{b^2 - d^2}{d^2 + b^2}$, and $\frac{c^2 - b^2}{c^2 + b^2}$ reſpectively,

muſt

muſt now become Ma^2, Mb^2, Mc^2, $M \times \overline{b^2 - c^2}$, $M \times \overline{c^2 - a^2}$, $M \times \overline{a^2 - b^2}$, $\frac{b^2 - c^2}{a^2}$, $\frac{c^2 - a^2}{b^2}$, and $\frac{a^2 - b^2}{c^2}$. And the three fundamental equations for the accelerative forces become $\frac{\overline{b^2 - c^2} \times yz}{a^2} = \frac{\dot{x}}{\dot{t}}$, $\frac{\overline{c^2 - a^2} \times zx}{b^2} = \frac{\dot{y}}{\dot{t}}$, and $\frac{\overline{a^2 - b^2} \times xy}{c^2} = \frac{\dot{z}}{\dot{t}}$, or $\dot{x} = \frac{\overline{b^2 - c^2} \times yz\dot{t}}{a^2}$, $\dot{y} = \frac{\overline{c^2 - a^2} \times zx\dot{t}}{b^2}$, $\dot{z} = \frac{\overline{a^2 - b^2} \times xy\dot{t}}{c^2}$; multiplying the firſt of theſe equations by a^2x, the ſecond by b^2y, and the third by c^2z, and adding all the three products or reſulting equations together gives $a^2x\dot{x} + b^2y\dot{y} + c^2z\dot{z} = 0$; alſo multiplying them reſpectively by a^4x, b^4y, and c^4z, and adding the three products produces $a^4x\dot{x} + b^4y\dot{y} + c^4z\dot{z} = 0$; and if $\mathfrak{A}$, $\mathfrak{B}$, and $\mathfrak{C}$, be the reſpective values of x, y, and z, at the commencement of the motion, taking the fluents $a^2x^2 + b^2y^2 + c^2z^2 = a^2\mathfrak{A}^2 + b^2\mathfrak{B}^2 + c^2\mathfrak{C}^2$, and $a^4x^2 + b^4y^2 + c^4z^2 = a^4\mathfrak{A}^2 + b^4\mathfrak{B}^2 + c^4\mathfrak{C}^2$, which therefore are conſtant quantities. But Ma^2x^2, Mb^2y^2, and Mc^2z^2, are the reſpective *vires vivæ* of the body round the three permanent axes, and conſequently their ſum, or the whole *vis viva* is always the ſame conſtant quantity. Alſo, ſince $\dot{t} = \frac{a^2\dot{x}}{\overline{b^2 - c^2} \times yz} = \frac{c^2\dot{z}}{\overline{a^2 - b^2} \times xy} = \frac{b^2\dot{y}}{\overline{c^2 - a^2} \, . \, xz}$, therefore $\frac{a^2x\dot{x}}{b^2 - c^2} = \frac{b^2y\dot{y}}{c^2 - a^2} = \frac{c^2z\dot{z}}{a^2 - b^2}$, and the fluents $\frac{a^2}{b^2 - c^2} \times \overline{x^2 - \mathfrak{A}^2} = \frac{b^2}{c^2 - a^2} \times \overline{y^2 - \mathfrak{B}^2} = \frac{c^2}{a^2 - b^2} \times \overline{z^2 - \mathfrak{C}^2}$; hence then $y = \sqrt{\frac{a^2 \times \overline{c^2 - a^2}}{b^2 \times \overline{b^2 - c^2}} \times \overline{x^2 - \mathfrak{A}^2} + \mathfrak{B}^2}$, and $z = \sqrt{\frac{a^2 \times \overline{a^2 - b^2}}{c^2 \times \overline{b^2 - c^2}} \times \overline{x^2 - \mathfrak{A}^2} + \mathfrak{C}^2}$, which values ſubſtituted for y and z in the equation $\dot{t} = \frac{a^2\dot{x}}{\overline{b^2 - c^2} \times yz}$, give $\dot{t}$ in terms of x, $\dot{x}$ and conſtant quantities. But the fluent, though attainable by means of the arcs of the conic ſections, is inſufficient for determining the motion of the body with reſpect to abſolute ſpace, becauſe at preſent nothing is found but the relations of *inertiæ* and velocities.

In order to determine a point which can be conſidered as at reſt in abſolute ſpace, and the nature of the body's motion with reſpect to it; let Z (fig. 8.) be ſuch a point, abſolutely at reſt itſelf, but ſo as to be always touched by the moving ſpherical ſurface which revolves with the body. Or, it is the ſame thing to conſider it as a given point upon a concave ſpherical ſurface at reſt, ſurrounding and every where touching that ſuppoſed above to revolve with the body. Through this point Z ſuppoſe quadrantal arcs A*l*, B*m*, and C*n*, to be drawn from the poles of the three permanent axes, and conſequently perpendicular to the three ſides of the octant ABC, ſuppoſing alſo Z to be at the inſtant over ſome point of this octant, and that *a* is greater than *b*, and *b* than *c*, when the velocity of the octant along its three ſides muſt neceſſarily be in the ſenſe from A towards B, from B towards C, and from C towards A; then (by ſpherics) as ſ. ZA : 1 :: ſ. Z*m* = coſ. ZB : ſ. ZAC = coſ. ZAB :: ſ. Z*n* = coſ. ZC : ſ. ZAB = coſ. ZAC; alſo, as ſ. BZ : 1 :: ſ. Z*n* = coſ. ZC · ſ. ZBA = coſ. ZBC :: ſ. Z*l* = coſ. ZA : ſ. ZBC = coſ. ZBA; and as ſ. CZ : 1 :: ſ. Z*l* = coſ. ZA : ſ. ZCB = coſ. ZCA :: ſ. Z*m* = coſ. BZ : ſ. ZCA = coſ. ZCB.

Now, the velocity z in AB reduced into the direction of the great circle ZA is $= z \times$ coſ. ZAB $= \frac{z \times \text{coſ. ZB}}{\text{ſ. ZA}}$, and the velocity y in the circle CA reduced into the direction of the great circle ZA $= y \times$ coſ. ZAC $= \frac{y \times \text{coſ. ZC}}{\text{ſ. ZA}}$, but in a contrary ſenſe to the former; conſequently the velocity of the point A along the great circle AZ in abſolute ſpace, that is, the velocity with which A approaches the

the fixed point Z muſt be $=\frac{z \times \text{coſ. ZB} - y \times \text{coſ. ZC}}{\text{ſ. ZA}}$; in like manner is found $\frac{x \times \text{coſ. ZC} - z \times \text{coſ. ZA}}{\text{ſ. ZB}}$ the velocity of B along BZ, and $\frac{y \times \text{coſ. ZA} - x \times \text{coſ. ZB}}{\text{ſ. ZC}}$ = that of C along CZ in abſolute ſpace. But the fluxions of the arcs ZA, ZB, and ZC, are $\frac{\text{coſ. Z}\dot{\text{A}}}{\text{ſ. ZA}}$, $\frac{\text{coſ. Z}\dot{\text{B}}}{\text{ſ. ZB}}$, and $\frac{\text{coſ. Z}\dot{\text{C}}}{\text{ſ. ZC}}$, reſpectively, which divided by their correſpondent velocities, give the fluxion of the time, that is, $\dot{t} = \frac{\text{coſ. Z}\dot{\text{A}}}{z \times \text{coſ. ZB} - y \times \text{coſ. ZC}} = \frac{a^2 \dot{x}}{zb^2 y - yc^2 z}$ (above found) $= \frac{\text{coſ. Z}\dot{\text{B}}}{x \times \text{coſ. ZC} - z \times \text{coſ. ZA}} = \frac{b^2 \dot{y}}{xc^2 z - za^2 x} = \frac{\text{coſ. Z}\dot{\text{C}}}{y \times \text{coſ. ZA} - x \times \text{coſ. ZB}} = \frac{c^2 \dot{z}}{ya^2 x - xb^2 y}$; from which ſix-fold equation, it is evident, by inſpection only, that if m = any conſtant quantity whatever, and $ma^2 x$ = coſ. ZA, $mb^2 y$ = coſ. ZB, and $mc^2 z$ = coſ. ZC, all the conditions thereof will be anſwered. Then, ſince coſ. ZA^2 + coſ. ZB^2 + coſ. ZC^2 = 1, its equal $m^2 a^4 x^2 + m^2 b^4 y^2 + m^2 c^4 z^2$ muſt alſo be = 1: but from the former part of the proceſs $a^4 x^2 + b^4 y^2 + c^4 z^2 = a^4 \mathfrak{A}^2 + b^4 \mathfrak{B}^2 + c^4 \mathfrak{C}^2$; therefore $m = \frac{1}{\sqrt{a^4 \mathfrak{A}^2 + b^4 \mathfrak{B}^2 + c^4 \mathfrak{C}^2}}$ a conſtant quantity; and ſ. AZ^2 = 1 − coſ. AZ^2 = coſ. BZ^2 + coſ. CZ^2 = $1 - m^2 a^4 x^2 = m^2 b^4 y^2 + m^2 c^4 z^2$, ſ. BZ^2 = 1 − coſ. BZ^2 = $1 - m^2 b^4 y^2 = m^2 a^4 x^2 + m^2 c^4 z^2$, and ſ. CZ^2 = $1 - m^2 c^4 z^2 = m^2 a^4 x^2 + m^2 b^4 y^2$; and, from above, the velocities with which A, B, and C, approach Z are reſpectively $\frac{\overline{b^2 - c^2} \times yz}{\sqrt{b^4 y^2 + c^4 z^2}}$, $\frac{\overline{c^2 - a^2} \times xz}{\sqrt{a^4 x^2 + c^4 z^2}}$, and $\frac{\overline{a^2 - b^2} \times xy}{\sqrt{a^4 x^2 + b^4 y^2}}$; but as a is ſuppoſed greater than c, $c^2 - a^2$ is negative, and the velocity therefore in a contrary ſenſe, conſequently the poles A and C muſt approach Z, whilſt B recedes from it. The reſpective velocities

ties of the points A, B, and C, in directions perpendicular to ZA, ZB, and ZC, being computed in like manner are $\frac{z\times\text{coſ. }ZC+y\times\text{coſ. }ZB}{\text{ſ. }ZA}$, $\frac{z\times\text{coſ. }ZC+x\times\text{coſ. }ZA}{\text{ſ. }ZB}$, and $\frac{x\times\text{coſ. }ZA+y\times\text{coſ. }ZB}{\text{ſ. }ZC}$, or $\frac{c^2z^2+b^2y^2}{\sqrt{b^4y^2+c^4z^2}}$, $\frac{c^2z^2+a^2x^2}{\sqrt{c^4z^2+a^4x^2}}$, and $\frac{a^2x^2+b^2y^2}{\sqrt{a^4x^2+b^4y^2}}$, and if each of the ſquares of theſe be added to each correſpondent ſquare of the three former, the reſulting ſums will be z^2+y^2, z^2+x^2, and x^2+y^2, which are the ſquares of the abſolute velocities of the poles A, B, and C, along their own proper tracks in abſolute ſpace, the operation thus proving itſelf. Hence we gain a clear idea of the motion of the body, during the time that the octant ABC takes in paſſing under Z, beginning at ſome point V in CB (or in AB as the caſe may happen) and ending at ſome point W in CA; that is, the point Z enters the octant when V touches Z, and quits it at W, the motion of the body or ſpherical ſurface that revolves with it under Z, being in the ſenſe from W towards V; that is, W approaching the fixed point Z whilſt V recedes from it. And ſince both the directions and velocities of the poles A, B, and C, in abſolute ſpace are given above, their tracks alſo may be determined by means of quadratures, as will be ſhewn hereafter. Again, the track VZW, on the moving ſpherical ſurface, which always paſſes under, or, ſome point of which, always touches Z as the body revolves; and the velocity with which it paſſes under it in every poſition may hence be determined. Thus, from the equation above found for the value of z, is eaſily obtained coſ. $CZ^2=m^2c^2z^2=\frac{c^2\times\overline{a^2-b^2}}{a^2\times\overline{b^2-c^2}}\times$ coſ. $AZ^2-\frac{m^2c^2a^2\times\overline{a^2-b^2}}{b^2-c^2}\times\mathfrak{A}^2+m^2c^4\mathfrak{C}^2$, the equation of the curve VZW upon the moving ſpherical ſurface, which will alſo be found to

be

be the equation of the curve when orthographically projected upon the plane of the great circle CA. For let the ſphere be thus projected, then the quadrants AB, BC (fig. 9.) will be projected into the right lines BA, BC, and if Z be the projected place of the fixed point at any inſtant, let fall the right line ZX perpendicular to BC; then, by the nature of the projection $ZX = \text{coſ. } AZ$, and $BX = \text{coſ. } CZ$, and if $\frac{a^2}{b^2-c^2} = A$, $\frac{b^2}{a^2-c^2} = B$, and $\frac{c^2}{a^2-b^2} = C$, the above equation becomes $BX^2 = \frac{c^4A}{a^4C} \times ZX^2 - \frac{m^2c^4A\mathfrak{A}^2}{C} + m^2c^4\mathfrak{C}^2$, and $ZX^2 = \frac{a^4C}{c^4A} \times (BX^2 + \frac{m^2c^4A\mathfrak{A}^2}{C} - m^2c^4\mathfrak{C}^2)$ the projected track therefore upon the plane is an hyperbola, whoſe centre is B, abſciſſa BX, and ordinate ZX, and taking $ZX = 0$, $BX = mc^2\sqrt{\mathfrak{C}^2 - \frac{A\mathfrak{A}^2}{C}}$ = the diſtance from B at which the curve cuts BC, and is therefore the ſemi-tranſverſe axis of the hyperbola. But this is only poſſible whilſt $C\mathfrak{C}^2$ is greater than $A\mathfrak{A}^2$; for if $C\mathfrak{C}^2 = A\mathfrak{A}^2$, $XZ = BX \times \frac{a^2}{c^2}\sqrt{\frac{C}{A}}$, the projected track is a right line BU, and the real one a great circle of the ſphere paſſing through B. If $A\mathfrak{A}^2$ be greater than $C\mathfrak{C}^2$ the track will no longer cut CB, but muſt cut BA, and BU will in both caſes be an aſymptote to the projected track. Since the track in all caſes croſſes the great circle CA, and we are at liberty to ſuppoſe the motion to begin at what point thereof we pleaſe, it may be ſuppoſed to commence where the track croſſes CA, and where, of conſequence, the velocity along CA is then $= 0$; we may therefore take the aſſumed quantity $\mathfrak{B} = 0$, and ſtill all the conditions of the problem be fulfilled, the expreſſions thus becoming more ſimple, for then $\frac{1}{m^2} = a^4\mathfrak{A}^2 + c^4\mathfrak{C}^2$ and $A\mathfrak{A}^2 - \frac{A \times \text{cos. } AZ^2}{m^2a^4} = \frac{B \times \text{cos. } BZ^2}{m^2b^4} = C\mathfrak{C}^2 - \frac{C \times \text{cos. } CZ^2}{m^2c^4}$.

Suppose W to be the point of CA and V that of CB which comes under Z; then at W cos. BZ = o, and cos. $AZ = ma^2\mathfrak{A}$ = s. CW; and at V, cos. AZ = o, and cos. $BZ = mb^2\mathfrak{A}\sqrt{\frac{A}{B}}$ = s. CV = s. CW $\times \frac{b^2}{a^2}\sqrt{\frac{A}{B}}$; CV and CW being a kind of semi-transverse and semiconjugate axes to the elliptic track on the spherical surface that passes under Z. And the gnomonical projection of the track on a plane touching the sphere at C, or the orthographical on the plane of the great circle BA (fig. 10.) becomes known; for here YZ = cos. BZ = CX; CY = XZ = cos. AZ, and $A\mathfrak{A}^2 - CY^2 \times \frac{A}{m^2a^4} = ZY^2 \times \frac{B}{m^2b^4}$ is the equation of the curve VZW which is the projection of the track on this plane, being an ellipsis whose semi-axes are s. CV and s. CW or $mb^2\mathfrak{A}\sqrt{\frac{A}{B}}$ and $ma^2\mathfrak{A}$, because $\frac{a^4B}{b^4A} \times ZY^2 = m^2a^4\mathfrak{A}^2 - CY^2$. Moreover, the perpendicular to the plane of the projection from Z on the plane to Z on the spherical surface itself = cos. $CZ = \sqrt{m^2c^4\mathfrak{C}^2 - \frac{c^4B}{b^4C} \times ZY^2} = \sqrt{1 - CZ^2} = \sqrt{1 - ZX^2 - ZY^2}$; and the fluxion of the track at Z upon the spherical surface $= \sqrt{\text{cos. } A\dot{Z}^2 + \text{cos. } B\dot{Z}^2 + \text{cos. } C\dot{Z}^2} = m\sqrt{a^4\dot{x}^2 + b^4\dot{y}^2 + c^4\dot{z}^2}$, and since $\dot{t} = \frac{A\dot{x}}{yz} = -\frac{B\dot{y}}{zx} = \frac{C\dot{z}}{xy}$, we thence obtain $\dot{x}^2 = \frac{B^2y^2\dot{y}^2}{A^2x^2}$, $\dot{z}^2 = \frac{B^2y^2\dot{y}^2}{C^2z^2}$, and the fluxion of the track $= m\dot{y}\sqrt{\frac{a^4B^2y^2}{A^2x^2} + b^4 + \frac{c^4B^2y^2}{C^2z^2}}$, which divided by $\dot{t}$ gives $m\sqrt{\frac{a^4y^2z^2}{A^2} + \frac{b^4z^2x^2}{B^2} + \frac{c^4y^2x^2}{C^2}}$ = the velocity with

with which the track paſſes under Z, but $z^2 = \mathfrak{C}^2 - \frac{B}{C} \times y^2$, and $x^2 = \mathfrak{A}^2 - \frac{By^2}{A}$, alſo $z^2x^2 = \frac{B^2y^4}{AC} - \overline{\frac{B\mathfrak{A}^2}{C} + \frac{B\mathfrak{C}^2}{A}} \times y^2 + \mathfrak{A}^2\mathfrak{C}^2$, which ſubſtituted for their equals give the velocity $= m\sqrt{-\frac{Ba^4y^4}{CA^2} + \frac{a^4y^2\mathfrak{C}^2}{A^2} + \frac{b^4y^4}{AC} - \frac{b^4y^2\mathfrak{A}^2}{BC} - \frac{b^4y^2\mathfrak{C}^2}{AB} + \frac{b^4\mathfrak{A}^2\mathfrak{C}^2}{B^2} - \frac{Bc^4y^4}{AC^2} + \frac{c^4y^2\mathfrak{A}^2}{C^2}} = m\sqrt{\frac{b^4\mathfrak{A}^2\mathfrak{C}^2}{B^2} - \frac{c^4y^2\mathfrak{C}^2}{AC} - \frac{a^4y^2\mathfrak{A}^2}{AC}} = \sqrt{\frac{m^2b^4\mathfrak{A}^2\mathfrak{C}^2}{B^2} - \frac{y^2}{AC}}$, becauſe $\frac{b^4}{AC} - \frac{Ba^4}{A^2C} - \frac{Bc^4}{C^2A} = \frac{B}{AC} \times \left(\frac{b^4}{B} - \frac{a^4}{A} - \frac{c^4}{C}\right) = \frac{B}{AC} \times (b^2 \times \overline{a^2 - c^2} - a^2 \times \overline{b^2 - c^2} - c^2 \times \overline{a^2 - b^2}) = 0$, $\frac{a^4}{A^2} - \frac{b^4}{AB} = -\frac{c^4}{AC}$ and $m^2a^4\mathfrak{A}^2 + m^2c^4\mathfrak{C}^2 = 1$. Now, ſuppoſing as above, the motion to begin when W is under Z and $y = 0$, the track muſt croſs CA at right angles, and with a velocity under $Z = \overline{a^2 - c^2} \times m\mathfrak{A}\mathfrak{C} = \frac{\overline{a^2 - c^2} \times \mathfrak{A}\mathfrak{C}}{\sqrt{a^4\mathfrak{A}^2 + c^4\mathfrak{C}^2}}$ that velocity being then the ſwifteſt poſſible, $\mathfrak{A}$, $\mathfrak{C}$, and $\sqrt{\mathfrak{A}^2 + \mathfrak{C}^2}$ being the then velocities of the poles C, A, and B, along their proper tracks in abſolute ſpace, the velocity x being then $= \mathfrak{A}$ and $z = \mathfrak{C}$, which are their greateſt values; and then Z becoming without the octant ABC, the velocity y muſt be negative or in a contrary ſenſe to what it would be if Z were within the octant; that is, ſince within the octant, y, as we have ſeen, is in the ſenſe from C towards A, it muſt now be in the ſenſe from A towards C; x and z ſtill continuing to be in the ſame ſenſe as if Z were within the octant, till the great circle BCV′ comes under Z which then touches V′, and conſequently $x = 0$, $y^2 = \frac{A\mathfrak{A}^2}{B}$, $z^2 = \mathfrak{C}^2 - \frac{A\mathfrak{A}^2}{C}$ and $\sqrt{\frac{m^2b^4\mathfrak{A}^2\mathfrak{C}^2}{B^2} - \frac{\mathfrak{A}^2}{BC}} = \frac{ma^2\mathfrak{A}}{\sqrt{ABC}}\sqrt{C\mathfrak{C}^2 - A\mathfrak{A}^2}$ = the velocity of the track under Z, which is then the ſloweſt, the correſpondent velocities of the poles A, B, and C, along their

their own proper tracks in abſolute ſpace being then $\sqrt{\mathfrak{C}^2 + \frac{A\mathfrak{A}^2}{B} - \frac{A\mathfrak{A}^2}{C}}$, $\sqrt{\mathfrak{C}^2 - \frac{A\mathfrak{A}^2}{C}}$, and $\mathfrak{A}\sqrt{\frac{A}{B}}$. And when V′ with the above found velocity has paſſed under Z, then the velocity x becomes negative; therefore, whilſt the point Z is within the angle formed by AC and BC produced beyond C both y and x are negative, till the great circle BC again croſſing under Z at W′, y is again $=0$, and the velocity of the track under Z the ſame as when W was under it, the correſponding velocities of the poles of the permanent axes being the ſame alſo; after which y will again become poſitive, x ſtill continuing negative during the time that Z is within the angle BCW′, till it again croſſes BC at V, and x is again $=0$, and the velocities of the track and permanent poles the ſame as when V′ croſſed under Z; afterwards the point Z being within the octant ABC, the velocities x, y, and z,, will be all poſitive till W again comes under Z, and another revolution under Z begins, *and ſo on for ever.* Moreover, the track being ſuppoſed to croſs CA and CB, when either W or W′ is under Z, the velocity $\sqrt{\mathfrak{A}^2 + \mathfrak{C}^2}$ of the pole B is the greateſt poſſible, being then $=$ the greateſt velocity that the ſpherical ſurface any where has or can have; and when V and V′ are under Z, $\sqrt{\mathfrak{C}^2 + \frac{A\mathfrak{A}^2}{B} - \frac{A\mathfrak{A}^2}{C}}$ $=$ the velocity of the pole A is the ſwifteſt which it can have, being then $=$ the greateſt *velocity* which the ſpherical ſurface any where has at that inſtant, ſuch *velocity* of the ſurface being then the *leaſt* poſſible.

Moreover, ſuppoſing ſtill the motion to begin when $y=0$, and $\mathfrak{B}=0$, $\dot{t} = -\frac{B\dot{y}}{xz} = -\frac{B\dot{y}}{\sqrt{\mathfrak{A}^2 - \frac{By^2}{A}}\sqrt{\mathfrak{C}^2 - \frac{By^2}{C}}} = -\frac{\dot{y}\sqrt{AC}}{\sqrt{\frac{A\mathfrak{A}}{B} - y^2}\sqrt{\frac{C\mathfrak{C}^2}{B} - y^2}}$;

let

let $y^2 = \frac{ACu}{B}$ or $y = u^{\frac{1}{2}}\sqrt{\frac{AC}{B}}$, $\dot{y} = \frac{1}{2}\dot{u}\sqrt{\frac{AC}{Bu}}$, and $\dot{t} = \frac{AC}{\sqrt{B}} \times \frac{-\dot{u}}{2u^{\frac{1}{2}}\sqrt{\frac{A\mathfrak{A}^2}{B} - \frac{ACu}{B}}\sqrt{\frac{C\mathfrak{C}^2}{B} - \frac{ACu}{B}}} = \frac{B^{\frac{1}{2}}}{2} \times \frac{-\dot{u}}{u^{\frac{1}{2}}\sqrt{\frac{\mathfrak{A}^2}{C} - u}\sqrt{\frac{\mathfrak{C}^2}{A} - u}}$; which here naturally divides into three forms or cafes, 1ſt, $\frac{B^{\frac{1}{2}}}{2} \times \frac{-\dot{u}}{\sqrt{\frac{\mathfrak{A}^2 u}{C} - u^2}\sqrt{\frac{\mathfrak{C}^2}{A} - u}}$; 2dly, $\frac{B^{\frac{1}{2}}}{2} \times \frac{-\dot{u}}{\sqrt{\frac{\mathfrak{C}^2 u}{A} - u^2}\sqrt{\frac{\mathfrak{A}^2}{C} - u}}$; 3dly, when $\frac{\mathfrak{A}^2}{C} = \frac{\mathfrak{C}^2}{A}$, it is $\frac{B^{\frac{1}{2}}}{2} \times \frac{-\dot{u}}{\frac{\mathfrak{A}^2 u^{\frac{1}{2}}}{C} - u^{\frac{3}{2}}} = -\frac{\dot{y}\sqrt{AC}}{\frac{A\mathfrak{A}^2}{B} - y^2}$; which laſt is of an eaſy and known form; and the fluents of the two former may be found by help of the arcs of the conic ſections; or otherwiſe, by the following contrivance.

Suppoſe a bar of metal, or other ſuch like body, whoſe centre of oſcillation is H (fig. 11.) to revolve at the earth's ſurface in a vertical plane without reſiſtance about the centre C, and that it is impelled from the loweſt point S with a velocity equal to that which would be acquired by an heavy body in falling freely by the force of uniform gravity through the height k, that is, if $2g$ = the force of gravity, ſuppoſe it impelled from S with a velocity $2\sqrt{gk}$ up the ſemicircle SMH, whoſe radius CS = CH = CM = r; then, MV being parallel to the horizon, and SV = u; its velocity at M muſt be $2\sqrt{gk - gu}$, and the fluxion of the arch MS = $\dot{MH} = \frac{-r \times \dot{HV}}{MV} = \frac{r\dot{u}}{\sqrt{2ru - u^2}}$, and the time of deſcribing $\dot{SM} = \frac{r}{2} \times \frac{-\dot{u}}{g^{\frac{1}{2}}\sqrt{2ru - u^2}\sqrt{k - u}}$ becauſe the velocity diminiſhes as SV increaſes, this fluxion compared with

with $\dot{t} = \frac{B^{\frac{1}{2}}}{2} \times \frac{-\dot{u}}{\sqrt{\frac{\mathfrak{A}^2 u}{C} - u^2}\sqrt{\frac{\mathfrak{C}^2}{A} - u}}$, we have $2r = \frac{C}{\mathfrak{A}^2}$, $k = \frac{\ }{C^2}$; if therefore $\frac{\mathfrak{A}^2}{2C} = CH$, we have, as $\frac{\mathfrak{A}^2}{2g^{\frac{1}{2}}C}$: the fluxion of the time of the bar's defcribing SM :: $B^{\frac{1}{2}} : \dot{t}$, that is, $\frac{\mathfrak{A}^2}{2C} : \surd Bg ::$ $\frac{-r\dot{u}}{2g^{\frac{1}{2}}\sqrt{2ru - u^2}\sqrt{k - u}} : \dot{t}$; but the velocity at $H = 2g^{\frac{1}{2}}\sqrt{k - 2r} =$ $2g^{\frac{1}{2}}\sqrt{\frac{\mathfrak{C}^2}{A} - \frac{\mathfrak{A}^2}{C}}$, if therefore $\frac{\mathfrak{C}^2}{A}$ be greater than $\frac{\mathfrak{A}^2}{C}$ (which may be called the firft cafe) the bar will make whole revolutions round the centre C, and its velocity at H = that acquired by an heavy body in falling through the height $\sqrt{\frac{\mathfrak{C}^2}{A} - \frac{\mathfrak{A}^2}{C}}$, and at S the arch MH = the femicircle. Now, when $y = 0$, that is, when W or W′ is under Z, $u = 0$, SV = 0, and when $u = 2r =$ $\frac{\mathfrak{A}^2}{C} = SH$, then $y^2 = \frac{A\mathfrak{A}^2}{B}$ which is the value of y^2 at V and V′ above, the afcent therefore of the bar from S to H in the femicircle correfponds to the motion of the body during the time that the quadrant of the track beginning at W and ending at V′ paffes under Z, and the fluxions of the times being to one another as $\frac{\mathfrak{A}^2}{2Cg^{\frac{1}{2}}} : B^{\frac{1}{2}}$, the times muft be in the fame ratio, confequently, as $\frac{\mathfrak{A}^2}{2C} : \surd Bg$ :: the time of two revolutions of the bar : the time of one revolution of the track WV′W′V under Z.

But if, as in cafe fecond, $\frac{\mathfrak{A}^2}{C}$ be greater than $\frac{\mathfrak{C}^2}{A}$, and r be ftill $= \frac{\mathfrak{A}^2}{2C}$ the bar can proceed no higher than till k = that height = $\frac{\mathfrak{C}^2}{A}$, its velocity at S being $= 2g\sqrt{\frac{\mathfrak{C}^2}{A}}$, when $u = 0$ and y and SV $= 0$;

$z=0$; and when $u=\frac{\mathfrak{C}^2}{A}$, $y^2=\frac{C\mathfrak{C}^2}{B}$ which is its value when $z=0$, as it ought to be, the track in this case, that is, when $A\mathfrak{A}^2$ is greater than $C\mathfrak{C}^2$, crossing AC and AB; the bar in this case making only oscillations and not revolutions. But if r now be made $=\frac{\mathfrak{C}^2}{2A}$ instead of $\frac{\mathfrak{A}^2}{2C}$, the bar will still make whole revolutions and as $\frac{\mathfrak{C}^2}{2A} : \sqrt{Bg}$ :: the time of two whole revolutions of the bar whose centre of oscillation is at $\frac{\mathfrak{C}^2}{2A}$ distance from C : the time of one revolution of the body under Z.

These cases may be otherwise resolved by finding the length $SC=r$, such that the bar may make two revolutions or oscillations whilst the body makes one; thus, let SV, instead of being $=u$, be in a constant ratio to it, or $SV = lu$, and $\dot{t}=\frac{r}{2}\times\frac{l\dot{u}}{g^{\frac{1}{2}}\sqrt{2rlu-l^2u^2}\sqrt{k-lu}}=\frac{B^{\frac{1}{2}}}{2}\times\frac{\dot{u}}{\sqrt{\frac{\mathfrak{A}^2u}{C}-u^2}\sqrt{\frac{\mathfrak{C}^2}{A}-u}}=\frac{B^{\frac{1}{2}}}{2}\times$ $\frac{\dot{u}l\sqrt{l}}{\sqrt{\frac{\mathfrak{A}^2l^2u}{C}-l^2u^2}\sqrt{\frac{\mathfrak{C}^2l}{A}-lu}}$, and comparing the homologous quantities, $\frac{l\sqrt{lB}}{2}=\frac{rl}{2g^{\frac{1}{2}}}$, $r=\sqrt{lBg}$, $2rl=\frac{\mathfrak{A}^2l^2}{C}$, $r=\frac{\mathfrak{A}^2l}{2C}=\sqrt{lBg}$, $\sqrt{l}=\frac{2C\sqrt{Bg}}{\mathfrak{A}^2}$, $l=\frac{4C^2Bg}{\mathfrak{A}^4}$, $r=\frac{2CBg}{\mathfrak{A}^2}$, $k=\frac{\mathfrak{C}^2l}{A}=\frac{4C^2Bg\mathfrak{C}^2}{A\mathfrak{A}^4}$; now, when such a bar makes whole revolutions, k must be greater than $2r$, or $\frac{4C^2Bg\mathfrak{C}^2}{A\mathfrak{A}^4}$ than $\frac{4CBg}{\mathfrak{A}^2}$, $\frac{C\mathfrak{C}^2}{A\mathfrak{A}^2}$ than unity, and $C\mathfrak{C}^2$ than $A\mathfrak{A}^2$. A bar therefore whose centre of oscillation is $\frac{2CBg}{\mathfrak{A}^2}$ distant from the centre of motion, will make two whole revolutions whilst the whole track WV′W′V moves once under Z if if $C\mathfrak{C}^2$ be greater than $A\mathfrak{A}^2$; but if $C\mathfrak{C}^2$ be lefs than $A\mathfrak{A}^2$ it will make two whole ofcillations. In like manner it will be found, that if $SC = r = \frac{2ABg}{\mathfrak{C}^2}$, fuch bar will make whole revolutions when $A\mathfrak{A}^2$ is greater than $C\mathfrak{C}^2$, and ofcillations when $A\mathfrak{A}^2$ is lefs than $C\mathfrak{C}^2$; and we are at liberty to make either the one fuppofition or the other.

Cafe 3. But if $A\mathfrak{A}^2 = C\mathfrak{C}^2$, and the track that paffes under Z be a great circle of the fphere, then $Ax^2 = Cz^2$, $\frac{x^2}{C} = \frac{z^2}{A}$, $By^2 = A\mathfrak{A}^2 - Ax^2 = A\mathfrak{A}^2 - Cz^2$, $\frac{y^2}{AC} = \frac{\mathfrak{A}^2 - x^2}{BC}$, and the velocity under $\mathfrak{Z} =$ $m\sqrt{\frac{b^4\mathfrak{A}^2\mathfrak{C}^2}{B^2} - c^4\mathfrak{C}^2 + a\,\mathfrak{A}^2 \times \frac{\mathfrak{A}^2 - x^2}{BC}} = \frac{m}{BC}\sqrt{b^4AC\mathfrak{A}^4 - \overline{\mathfrak{A}^2 - x^2} \times \overline{c^4AB\mathfrak{A}^2 + a\,BC\mathfrak{A}^2}} = \frac{m\mathfrak{A}x}{BC}\sqrt{c^4AB + a^4BC} = \frac{m\mathfrak{A}b^2x}{B}\sqrt{\frac{A}{C}}$, which is therefore $= 0$ when $x = 0$, or B is under Z, fuppofing that to be poffible. But then $\dot{t} = \frac{-\dot{y}\sqrt{AC}}{\frac{A\mathfrak{A}^2}{B} - y^2} = \frac{\sqrt{CB}}{2\mathfrak{A}} \times 2\mathfrak{A}\sqrt{\frac{A}{B}} \times \frac{-\dot{y}}{\frac{A\mathfrak{A}^2}{B} - y^2}$, and $t = \frac{\sqrt{CB}}{2\mathfrak{A}}$

$\times$ hyp. log. of $\frac{\mathfrak{A}\sqrt{\frac{A}{B}} - y}{\mathfrak{A}\sqrt{\frac{A}{B}} + y}$; therefore, when at the firft inftant $y = 0$, to have the motion poffible, y muft be a negative quantity; which is agreeable to what was obferved before, that y muft be negative within the angle ACV′; but in this cafe Z can never come over V′, for then t would be infinite. And if the motion be fuppofed to begin when Z is fomewhere within the octant ABC, where y the firft inftant is equal to a given quantity $\mathfrak{B}$, then the fluent muft be fo corrected as that $t = \frac{\sqrt{CB}}{2\mathfrak{A}}$ × hyp. log. of

$\frac{\mathfrak{A}\sqrt{\frac{A}{B}}+\mathfrak{B}}{\mathfrak{A}\sqrt{\frac{A}{B}}-\mathfrak{B}} \times \frac{\mathfrak{A}\sqrt{\frac{A}{B}}-y}{\mathfrak{A}\sqrt{\frac{A}{B}}+y}$, but the time or motion can never begin at B, nor can the pole oppoſite to B ever come under Z. And the reaſon of this is alſo evident from the nature of the motion itſelf; for theſe being poles of a permanent axis, if Z were once over one of them, it muſt always continue ſo.

Having thus determined the time, velocity, and manner, in which the ſpherical ſurface that revolves with the body paſſes under the fixed point Z, it only remains to determine the path of one of the poles as C of the permanent axes about Z in abſolute ſpace, or upon a ſpherical ſurface at reſt, but equal and concentric with that ſuppoſed to move with the body; for the path of one of theſe poles as C being found, thoſe of the other two, and indeed the path of every other invariable point of the moving ſpherical ſurface, becomes known.

Now, the velocity with which C approaches Z is found above $= \frac{\overline{a^2-b^2} \times xy}{\sqrt{a^4x^2+b^4y^2}}$, and the fluxion of the arc $CZ = \frac{\text{ſ. } C\dot{Z}}{\text{coſ. } CZ} = -\frac{\text{coſ. } C\dot{Z}}{\text{ſ. } CZ}$ divided by the velocity gives $\dot{t}$, whoſe fluent is found above, and conſequently the diſtance of C from Z at the end of any time t, there is then only wanting the angle deſcribed by C about Z, correſponding to the diſtance CZ therefrom, to have the path of C about Z; which may be found by the help of quadratures as follows.

As ſ. ZC : velocity of C perpendicular to ZC (found above) $= \frac{x \times \text{coſ. } AZ + y \times \text{coſ. } BZ}{\text{ſ. } ZC}$ :: 1 : the angular velocity of C about $Z = \frac{x \times \text{coſ. } AZ + y \times \text{coſ. } BZ}{\text{ſ. } ZC^2}$ which velocity being multiplied by

coſ.

$\frac{\text{cof. } Z\dot{C}}{y \times \text{cof. } AZ - x \times \text{cof. } BZ}$ the fluxion of the time, gives the fluxion of the angle defcribed by C about $Z = \frac{\text{cof.} Z\dot{C}}{\text{f. } ZC^2} \times \frac{x \times \text{cof. } AZ + y \times \text{cof. } BZ}{y \times \text{cof. } AZ - x \times \text{cof. } BZ}$ $= \frac{\text{cof. } Z\dot{C}}{\text{f. } ZC^2} \times \frac{b^2 \times \text{cof. } AZ^2 + a^2 \times \text{cof. } ZB^2}{a^2 - b^2 \times \text{cof. } ZA \times \text{cof. } ZB}$, which in terms of ZC is by computation $= \frac{\text{cof. } Z\dot{C}}{\text{f. } ZC^2} \times \frac{ba^2 \times \text{f. } ZC^2 - b \times \overline{a^2 - c^2} \times s^2}{a\sqrt{\overline{b^2 - c^2} \times \overline{a^2 - c^2}}\sqrt{\text{f.} CV^2 - \text{f.} ZC}\sqrt{n^2 - \text{cof.} ZC^2}}$; where s and $n =$ the fine and cofine of CW, and $\text{f. } CV^2 = \frac{b^2 s^2}{a^2} \times \frac{a^2 - c^2}{b^2 - c^2}$. Now, this being the fluxion of the arc to radius 1, which is the meafure of the angle defcribed by C about Z in the time t; this arc in value therefore will be double the area of the fector of the circle whofe radius is unity defcribed about Z in the fame time. Hence, having found a fector of a circle to radius unity, whofe area is half the fluent of the above fluxion, or the fluent of half the above fluxion, the arch-line of this fector will be the meafure of the required angle defcribed by C about Z in the time t.

Let $\dot{A} = \frac{\text{cof. } Z\dot{C}}{\sqrt{n^2 - \text{cof. } ZC^2}}$, A being the arc, beginning when $n =$ cof. ZC, whofe cofine $= \frac{\text{cof. } ZC}{n}$ and radius unity, and $\dot{B} = \frac{-\text{f. } Z\dot{C}}{\sqrt{\text{f. } CV^2 - \text{f. } ZC^2}}$, B being = the arc, beginning when CV = ZC, whofe cofine $= \frac{\text{f. } ZC}{\text{f. } CV}$ and radius unity, and in fig. 12. take ZY fuch that $\frac{\text{cof. } Z\dot{C}}{\sqrt{n^2 - \text{cof. } ZC^2}} \times \frac{ba^2 \times \text{f. } ZC^2 - b \times \overline{a^2 - c^2} \times s^2}{2a\sqrt{\overline{b^2 - c^2} \times \overline{a^2 - c^2}} \times \text{f.} ZC^2 \sqrt{\text{f.} CV^2 - \text{f.} ZC^2}}$ may $= \dot{A} \times \frac{ZY^2}{2}$ the fluxion of the curvilinear area defcribed about the centre Z and bounded by the ordinate ZY, whofe firft value is ZG when $n =$ cof. ZC, and $A = 0$; on ZG take

ZS = 1, with which radius on the centre Z deſcribe the circle STR′ on which take ST = any value of A, and through T draw ZY = the ordinate correſponding to that value of A, and thus may points at pleaſure be found, and the curve GY conſtructed. Now, when ZC = CV, the value of ZY = $\sqrt{\frac{ba^2 \times \text{ſ. ZC}^2 - bs^2 \times \overline{a^2 - c^2}}{a\sqrt{\overline{b^2 - c^2} \times \overline{a^2 - c^2}} \times \text{ſ. ZC}^2 \sqrt{\text{ſ.CV}^4 - \text{ſ.ZC}^4}}}$ is infinite, and if SR = the then value of the arc A, ZR produced will be an aſymptote to the curve GY. But to remedy this inconveniency ariſing to the conſtruction from this infinite length of the curve; produce any other radius ZR′ of the circle, till ZH = the firſt value of ZY′ $\sqrt{\frac{ba^2 \times \text{ſ. ZC}^2 - bs^2 \times \overline{a^2 - c^2}}{a\sqrt{\overline{b^2 - c^2} \times \overline{a^2 - c^2}} \times \text{ſ. ZC} \times \text{coſ. ZC} \sqrt{n^2 - \text{coſ. CZ}^2}}}$, when CV = ZC and the arc B = o, and taking ZY′ = any other value thereof correſponding to ſome value R′T′ of the arc B leſs than RS′ the value thereof when n = coſ. ZC and ZY′ infinite; and thus the curve HY may be conſtructed by points; let the conſtructions of both theſe curves GY and HY′ be continued till the value of the arc ZC in the one conſtruction be equal to that in the other; then muſt the ſum of the correſponding areas ZGY + ZHY′ be equal to the infinitely extended area formed by each curve running out towards its own aſymptote, each of theſe infinitely extended areas being equal becauſe they begin together, and are the fluents of the equal fluxions $\dot{A} \times \frac{ZY^2}{2}$ and $\dot{B} \times \frac{ZY'^2}{2}$. Equal to any value of the area ZGY, let the ſector QZR be cut off from the circle whoſe radius is unity; then the area of this ſector = half the arc RQ = the fluent of $\frac{\text{coſ. Z}\dot{\text{C}}}{\sqrt{n^2 - \text{coſ. ZC}^2}} \times \frac{ba^2 \times \text{ſ. ZC}^2 - b s^2 - \overline{a^2 - c^2}}{2a\sqrt{\overline{b^2 - c^2} \times \overline{a^2 - c^2}} \times \text{ſ.ZC}^2 \sqrt{\text{ſ.CV}^2 - \text{ſ.ZC}^2}}$;

 and

and the fluent of $\frac{-\text{ſ. } Z\dot{C}}{\sqrt{\text{ſ. } CV^2 - \text{ſ. } ZC^2}} \times$ $\frac{ba^2 \times \text{ſ. } ZC^2 - bs^2 \times \overline{a^2 - c^2}}{2a\sqrt{\overline{b^2 - c^2} \times \overline{a^2 - c^2}} \times \text{ſ. } ZC \times \text{coſ. } ZC \sqrt{n^2 - \text{coſ. } ZC^2}}$ alſo = the ſector QZV = ½VQ = the fluent of $\dot{B} \times \frac{ZY'^2}{2}$, the former being that of $\dot{A} \times \frac{ZY^2}{2}$. Then, ſuppoſing ſtill the motion to begin when $y = 0$, or ZC = CW, the arch QR muſt be the meaſure of the angle deſcribed by C about Z in the time *t*; and the whole arch RQV = the meaſure of the angle deſcribed during the time that ZC from being = CW becomes = CV, that is, during one-fourth of the time in which the track on the ſpherical ſurface makes one revolution or paſſes once under Z. Conſequently, if on ZR there be taken the right line ZC = the ſine of CW, and on CV, ZC″ = ſ. CV, and upon the intermediate radii as ZQ their correſpondent values of ſ. ZC, a curve drawn through all theſe points C, C′, C″, &c. will be the orthographical projection (upon a plane 90° from Z) of that which is the locus of C in abſolute ſpace, or upon the immoveable ſpherical ſurface; ſuch locus touching the circle whoſe radius ZC = ſ. CW at C, and that whoſe radius ZC″ = ſ. CV at C″. And the time of moving from C where ZC = ſ. CW to C″ where ZC″ = ſ. CV will be equal to that of a ſemirevolution a ſemivibration of the bar above found; and every ſucceeding part of the curve as C″, C‴, C⁗, deſcribed in the ſame or an equal time will be perfectly equal and ſimilar to C, C′, C″. If the angle CZC⁗ be a diviſor of 360°, the path will return into itſelf; if not, it will croſs itſelf ſomewhere as at C^{v}, and ſo on for ever.

GENERAL

GENERAL SCHOLIA.

1. Since the moving ſpherical ſurface paſſes under the fixed point Z in the ſenſe from Z towards V, and the invariable pole or point C on that ſurface moves round Z in a contrary ſenſe BCA (fig. 4. and 8.) there muſt be ſome point as O upon the ſurface which muſt be at reſt with reſpect to both theſe motions, and which point O muſt be the pole of the momentary axis, as will appear preſently; for the preceding ſolution being completed without any regard to ſuch axis, it may now be proper to deduce the properties of this axis therefrom, as by theſe means ſome *new light* may ſtill be caſt upon the motion under conſideration.

Let O (fig. 4.) be ſuch an axis, whoſe properties are conſidered in the propoſitions preceding the laſt, and let the angular velocity of the body about it $=\varpi$, coſ. AO$=\beta$, coſ BO$=\gamma$, coſ. CO$=\delta$; then it has been already ſhewn, that $\varpi\beta=x$, $\varpi\gamma=y$, and $\varpi\delta=z$; let theſe values be ſubſtituted for x, y, and z, in the general equations of the laſt propoſition; then $\beta^2+\gamma^2+\delta^2=1$, $x^2+y^2+z^2=\varpi^2=\varpi^2\beta^2+\varpi^2\gamma^2+\varpi^2\delta^2$, and ſuppoſing ſtill the motion to begin when $y=0$, $\gamma=0$, and $\varpi^2=x^2+z^2=\mathfrak{A}^2+\mathfrak{C}^2=e^2$; that is, let $e=$ the angular velocity about the momentary axis when its pole O croſſes the great circle AC; then, ſince $x^2=\mathfrak{A}^2-\frac{B}{A}\times y^2$, and $z^2=\mathfrak{C}^2-\frac{By^2}{C}$, $x^2+y^2+z^2$ $=\varpi^2=\mathfrak{A}^2-\frac{By^2}{A}+\mathfrak{C}^2-\frac{By^2}{C}+y^2=e^2+\varpi^2\gamma^2\times\overline{1-\frac{B}{A}-\frac{B}{C}}=e^2-\frac{\varpi^2\gamma^2}{AC}$ (becauſe $1-\frac{B}{A}-\frac{B}{C}=-\frac{1}{AC}$), and $\varpi^2=\frac{e^2}{1+\frac{\gamma^2}{AC}}$, which therefore can never be conſtant whilſt γ or BO is variable, except either $\frac{1}{A}$ or $\frac{1}{C} = 0$, that is, when either $b^2 = c^2$ or $a^2 = b^2$. In like manner it will alſo be found, that $\varkappa^2 = \frac{e^2 - \frac{\mathfrak{C}^2}{BA}}{1 - \frac{\delta^2}{BA}} = \frac{e^2 - \frac{\mathfrak{A}^2}{BC}}{1 - \frac{\beta^2}{BC}}$; and $\delta^2 = \frac{BA}{e^2 - \frac{\mathfrak{A}^2}{BC}} \times \left(\frac{\mathfrak{C}^2}{BA} - \frac{\mathfrak{A}^2}{BC} + \overline{e^2 - \frac{\mathfrak{C}^2}{BA}} \times \frac{\beta^2}{BC}\right)$, and when $\beta = 0$, or the pole of the momentary axis croſſes BC, $\delta^2 = \frac{BA \times \overline{\frac{\mathfrak{C}^2}{BA} - \frac{\mathfrak{A}^2}{BC}}}{e^2 - \frac{\mathfrak{A}^2}{BC}} = \frac{\mathfrak{C}^2 - \frac{A\mathfrak{A}^2}{C}}{e^2 - \frac{\mathfrak{A}^2}{BC}}$, and to have this poſſible it is neceſſary that $C\mathfrak{C}^2$ be greater than $A\mathfrak{A}^2$, and it is above determined, that under the ſame limitation Z muſt alſo croſs BC.

Again, from the equation $\frac{e^2}{1 + \frac{\gamma^2}{AC}} = \frac{e^2 - \frac{\mathfrak{C}^2}{BA}}{1 - \frac{\delta^2}{BA}}$, $e^2 - \frac{e^2\delta^2}{BA} = e^2 - \frac{\mathfrak{C}^2}{BA} + \frac{e^2\gamma^2}{AC} - \frac{\mathfrak{C}^2\gamma^2}{BCA^2}$, $\delta^2 = \frac{\mathfrak{C}^2}{e^2} + \frac{\mathfrak{C}^2\gamma^2}{ACe^2} - \frac{B\gamma^2}{C}$, and when $\gamma = 0$, or O croſſes CA, $\delta^2 = \frac{\mathfrak{C}^2}{e^2}$, let $\frac{\mathfrak{C}}{e} = m$, and then $\delta^2 = m^2 + \frac{m^2\gamma^2}{AC} - \frac{B\gamma^2}{C}$, or $m^2 - \delta^2 = \frac{B\gamma^2}{C} - \frac{m^2\gamma^2}{AC}$, which is the very equation brought out by a different method in the firſt ſcholium to the ſixth propoſition above. And if $n = \frac{\mathfrak{A}}{e} =$ the coſ. of the arc of which m is the ſine, it will be found in the very ſame manner that $\beta^2 = n^2 + \frac{n^2\gamma^2}{AC} - \frac{B\gamma^2}{A}$. Moreover, becauſe $A \times \overline{\mathfrak{A}^2 - x^2} = By^2 = C \times \overline{\mathfrak{C}^2 - z^2} = A \times \overline{\mathfrak{A}^2 - \varkappa^2\beta^2} = B\varkappa^2\gamma^2 = C \times \overline{\mathfrak{C}^2 - \varkappa^2\delta^2}$, $\varkappa^2 = \frac{A\mathfrak{A}^2}{B\gamma^2 + A\beta^2} = \frac{C\mathfrak{C}^2}{B\gamma^2 + C\delta^2}$, $\frac{A\mathfrak{A}^2}{C\mathfrak{C}^2} = \frac{An^2}{Cm^2} = \frac{B\gamma^2 + A\beta^2}{B\gamma^2 + C\delta^2}$, $\dot{t} = -\frac{B\dot{y}}{xz} = -B \times \frac{\varkappa\dot{\gamma} + \gamma\dot{\varkappa}}{\varkappa^2\beta\delta} = -B \times \frac{\varkappa^2\gamma\dot{\gamma} + \gamma^2\varkappa\dot{\varkappa}}{\varkappa^3\beta\gamma\delta}$, but $\varkappa^2 \times \overline{1 + \frac{\gamma^2}{AC}} = e^2$ a conſtant quantity; therefore $\varkappa\dot{\varkappa} \times \overline{1 + \frac{\gamma^2}{AC}} +$

$\varkappa^2$

$\frac{\vartheta^2\gamma\dot{\gamma}}{AC}=0$, and $\dot{t}=\frac{ABC\dot{\vartheta}}{\vartheta^2\beta\gamma\delta}$, or $\frac{\dot{\vartheta}}{\dot{t}}=\frac{\vartheta^2\beta\gamma\delta}{ABC}=$ the accelerating force acting along the midcircle at 90° from O. Since, when $\gamma=0$, $y=0$, and cos. BZ $=0$, the points Z and O are both upon CA at the same instant, and when $\beta=0$, $x=0$, and cos. AZ $=0$, also when $\delta=0$, $z=0$, and cos. CZ $=0$; therefore the poles Z and O both enter the octant ABC at the same instant; both, when $C\mathfrak{C}^2$ is greater than $A\mathfrak{A}^2$, cross BC at the same instant but at different points, *viz.* Z at V where s. CV² $=$ s. CW² $\times\frac{b^2}{a^2}\times\frac{a^2-c^2}{b^2-c^2}=\frac{\mathfrak{A}^2\times a^2b^2\times\overline{a^2-c^2}}{\overline{b^2-c^2}\times\overline{a^4\mathfrak{A}^2+c^4\mathfrak{C}^2}}$; and O where cos. BO² $=\gamma^2=$

$$\frac{n^2}{\frac{B}{A}-\frac{n^2}{AC}}=\frac{\mathfrak{A}^2}{\frac{B\mathfrak{C}^2}{A}+\mathfrak{A}^2-\frac{B\mathfrak{A}^2}{C}}=\frac{\mathfrak{A}^2a^2c^2\times\overline{a^2-c^2}}{\overline{b^2-c^2}\times\overline{b^2c^2\mathfrak{C}^2+a^2\mathfrak{A}^2\times\overline{b^2-c^2}\times\overline{b^2+c^2-a^2}}}$$

which cannot be greater than the corresponding value of s. CV² above; for, suppose the contrary, and that cos. BO² is greater than s.VC², then must $\frac{c^2}{b^2c^2\mathfrak{C}^2+a^2\mathfrak{A}^2\times\overline{b^2+c^2-a^2}}$ be greater than $\frac{b^2}{a^4\mathfrak{A}^2+c^4\mathfrak{C}^2}$ $c^2a^4\mathfrak{A}^2+c^6\mathfrak{C}^2$ than $b^4c^2\mathfrak{C}^2+b^2a^2\mathfrak{A}^2\times\overline{b^2+c^2-a^2}$; $\mathfrak{A}^2\times\overline{c^2a^4-a^2b^4-a^2b^2c^2+a^2b^2}$ than $\overline{b^4c^2-c^6}\times\mathfrak{C}^2$, $a^2\mathfrak{A}^2\times\overline{a^2-b^2}\times\overline{c^2+b^2}$ than $c^2\mathfrak{C}^2\times\overline{b^4-c^4}$; $a^2\mathfrak{A}^2\times\overline{a^2-b^2}$ than $c^2\mathfrak{C}^2\times\overline{b^2-c^2}$, and $A\mathfrak{A}^2$ than $C\mathfrak{C}^2$ which is impossible whilst Z crosses BC, because it has been proved, that then $C\mathfrak{C}^2$ is greater than $A\mathfrak{A}^2$; consequently O crosses BC between V and C (in fig. 8.) and both O and Z quit the octant ABC at the same instant; Z at W, and O between W and C, at the point where $\gamma^2=0$, and $\beta^2=n^2=\frac{\mathfrak{A}^2}{e^2}$, and, as will be more fully shewn, a great circle drawn from O to Z being always perpendicular to the track VZW. In the very same manner it may be shewn, that when $A\mathfrak{A}^2$ is greater than $C\mathfrak{C}^2$, and the track which passes under Z

2 crosses

croſſes BA, both O and Z ſtill enter the octant ABC together, both paſs over it in the ſame time, and both quit it or croſs CA together; but in this caſe the track for *Z* upon the moving ſurface is leſs than, or within, that of O, Z croſſing BA at a point nearer to A than that where O croſſes it; and O in both theſe caſes ſhifts its place on the moving ſpherical ſurface making one revolution in the time that the whole curve WV′W′V takes in paſſing under *Z*; both curves being ſuch that in the caſes above deſcibed where the projection of WV′W′V is a conic ſection, that of the track of O projected upon the ſame plane will be a conic ſection alſo, that is, where it is ſhewn above that the projecton of WV′W′V is an hyperbola, that of the track O will be an hyperbola, and an ellipſis where that of the other is an ellipſis.

And when $A\mathfrak{A}^2 = C\mathfrak{C}^2$ or $\mathfrak{C}^2 : \mathfrak{A}^2 :: A : C :: \delta^2 : \beta^2$, the track of O as well as Z is a great circle of the ſphere, ſince $\frac{\mathfrak{A}^2}{\mathfrak{C}^2} = \frac{C}{A}$, and ſ. $CQ^2 = \frac{C}{A} \times$ ſ. AQ^2 when O croſſes CA at Q (fig. 4.); and when *Z* croſſes CA coſ. $AZ^2 =$ ſ. $CW^2 = \frac{a^4\mathfrak{A}^2}{a^4\mathfrak{A}^2 + c^4\mathfrak{C}^2}$, and ſ. $CQ^2 = \frac{Cm^2}{A} = \frac{C}{A} \times \frac{\mathfrak{C}^2}{\mathfrak{A}^2 + \mathfrak{C}^2} = \frac{C}{C+A}$, and ſ. $CW^2 = \frac{1}{1 + \frac{c^4 A}{a^4 C}} = \frac{C}{C + \frac{a^4 A}{a^4}}$, conſequently, a^2 being, by hypotheſis, greater than c^2, the ſine of CW muſt be greater than ſ. CQ or than ſ. CO when O croſſes CA; and therefore the point where O croſſes CA muſt be nearer C than the point where *Z* croſſes it the ſame inſtant, in the caſe where both the tracks are great circles of the ſphere, paſſing through the ſame point B.

2. It is now well known, that the *momentum* of *inertia* of the body round the axis whoſe pole is O is $= Ma^2\beta^2 + Mb^2\gamma^2 + Mc^2\delta^2$, and if this be drawn into $\varkappa^2$, the product $M\varkappa^2 \times (a^2\beta^2 +$

b^2,

$b^2\gamma^2 + c^2\delta^2) = M \times (a^2x^2 + b^2y^2 + c^2z^2)$ = the whole *vis viva* of the body, or becaufe radius is unity, it is = the centrifugal motive force of the body round the natural or momentary axis, which being equal to the fum of Ma^2x^2, Mb^2y^2, and Mc^2z^2, thofe round the three permanent ones, and being above proved to be a conftant quantity, the perturbating motive forces $M \times \overline{b^2 - c^2} \times yz$, $M \times \overline{c^2 - a^2} \times zx$, and $M \times \overline{a^2 - b^2} = xy$, above found, cannot alter the *vis viva*, or whole motive force of the body along the midcircle, or that which is 90° from O. But, for a more particular proof of this, let thefe oblique perturbating motive forces be refolved into three others acting in the direction of the midcircle; the firft fo refolved being $= M \times \overline{b^2 - c^2} \times yz\beta = M\varkappa^2\beta\gamma\delta \times \overline{b^2 - c^2}$, the fecond $= M\varkappa^2\beta\gamma\delta \times \overline{c^2 - a^2}$, and the third $= M\varkappa^2\beta\gamma\delta \times \overline{a^2 - b^2}$; their fum $M\varkappa^2\beta\gamma\delta \times (b^2 - c^2 + c^2 - a^2 + a^2 - b^2)$ being = 0, fhews that there is no motive force in the direction of the midcircle arifing from them, wherefore that along the midcircle muft remain unaltered. But, though there is no perturbating motive force in the direction of the midcircle, there is neverthelefs an accelerative one acting along it; for the three perturbating accelerative forces round the three permanent axes being $\frac{b^2 - c^2}{a^2} \times yz$, $\frac{c^2 - a^2}{b^2} \times zx$, and $\frac{a^2 - b^2}{c^2} \times xy$, thefe being refolved into the direction of the midcircle, their fum $\varkappa^2\beta\gamma\delta \times \left(\frac{b^2 - c^2}{a^2} + \frac{c^2 - a^2}{b^2} + \frac{a^2 - b^2}{c^2}\right) = \varkappa^2\beta\gamma\delta \times \left(\frac{1}{A} - \frac{1}{B} + \frac{1}{C}\right)$ will not be = 0, but to $\frac{\varkappa^2\beta\gamma\delta}{ABC}$ which is the value of $\frac{u}{t}$ found in the preceding fcholium, and by the general properties of all fpherical motion as proved in the fourth propofition above is the accelerating force acting there.

This

This matter M. EULER confiders in a fomewhat different light, by finding the *initial axis*, or that about which, if the body were perfectly at reft, it would be firft urged to turn by accelerating forces acting upon it; and from Scholium 1. Prop. IV. above it appears, that if the body were at reft, and acted upon by three external accelerating forces $\frac{\dot{x}}{i}$, $\frac{\dot{y}}{i}$, and $\frac{\dot{z}}{i}$, it would be urged to turn the firft inftant about fome axis whofe pole is E by a fingle force $= \frac{\sqrt{\dot{x}^2+\dot{y}^2+\dot{z}^2}}{i}$, fuch that the five forces, $\frac{\sqrt{\dot{x}^2+\dot{y}^2+\dot{z}^2}}{i}$, $\frac{\dot{x}}{i}$, $\frac{\dot{y}}{i}$, $\frac{\dot{z}}{i}$ and $\frac{\dot{u}}{i}$ will be refpectively as radius, cof. EA, cof. EB, cof. EC, and cof. EO, or $\frac{\dot{x}}{i \times \text{cof. EA}} = \frac{\dot{y}}{i \times \text{cof. EB}} = \frac{\dot{z}}{i \times \text{cof. EC}} = \frac{\dot{u}}{i \times \text{cof. EO}}$, and fince when the body is in motion, and that motion difturbed by the unequal action of its own particles which generates accelerating forces, fuch forces confidered fimply in themfelves muft ftill have the fame tendency to turn the body about fome axis whofe pole is E different from that whofe pole is O, and fuch that the above equation may ftill obtain, and if the above-found values of $\frac{\dot{x}}{i}$, $\frac{\dot{y}}{i}$, and $\frac{\dot{z}}{i}$, be fubftituted therein, by means of a *calculus* fo inftituted, the value of $\dot{u}$, and confequently u will come out the very fame as by the preceding methods.

3. It ftill remains to be fhewn, that the point Z now determined has the properties fhewn to be requifite in the fifth Propofition above, *viz.* that it is at reft in abfolute fpace, and therefore at reft both with refpect to the motion of the fpherical furface, and to the velocity with which O the pole of the momentary axis fhifts its place. Now, by Scholium 1. Prop.

IV.

IV. the momentary pole ſhifts its place along its own track with a velocity $= \frac{\sqrt{\beta^2 + \dot{\gamma}^2 + \dot{\delta}^2}}{t}$; and if Z be ſuch as that the great circle OZ (fig. 4.) may always be perpendicular to the track WV′W′V (fig. 8.) that paſſes under Z, which it muſt be if O be the pole of motion; then as $1 : \varepsilon^2 ::$ ſ. $OZ^2 =$ coſ. $OY^2 :$ the ſquare of the velocity of the track under $Z = \frac{m^2 b^4 \mathfrak{A}^2 \mathfrak{C}^2}{B^2} - \frac{y^2}{AC}$, hence ſ. $OZ^2 = \frac{m^2 b^4 \mathfrak{A}^2 \mathfrak{C}^2}{B^2 \varepsilon^2} - \frac{y^2}{AC\varepsilon^2}$, coſ. $OZ^2 = 1 + \frac{y^2}{AC\varepsilon^2} - \frac{m^2 b^4 \mathfrak{A}^2 \mathfrak{C}^2}{B^2 \varepsilon^2} = 1 + \frac{\gamma^2}{AC} - \frac{m^2 b^4 \mathfrak{A}^2 \mathfrak{C}^2}{B^2 \varepsilon^2} = \frac{e^2}{\varepsilon^2} - \frac{m^2 b^4 \mathfrak{A}^2 \mathfrak{C}^2}{B^2 \varepsilon^2} = \frac{\overline{a^2 \mathfrak{A}^2 + c^2 \mathfrak{C}^2}|^2}{\varepsilon^2 \times \overline{a^4 \mathfrak{A}^2 + c^4 \mathfrak{C}^2}}$, and coſ. $ZO = \frac{a^2 \mathfrak{A}^2 + c^2 \mathfrak{C}^2}{\varepsilon \sqrt{a^4 \mathfrak{A}^2 + c^4 \mathfrak{C}^2}} =$ ſ. OY; this, then, is the value of coſ. ZO deduced from the ſuppoſition that it is always perpendicular to the track upon the moving ſpherical ſurface which paſſes under Z at reſt; and if this be found to agree with the value thereof computed by trigonometry, it will prove the legitimacy of that ſuppoſition, and that it is the true value ſuch as that O ſhall be always the pole of the momentary axis and Z at reſt in abſolute ſpace. Produce BZ (fig. 4.) till it cuts AC perpendicularly at q; then it is before found, that the ſine and coſine of BZ are $m\sqrt{a^4x^2 + c^4z^2}$ and mb^2y, thoſe of CZ $m\sqrt{a^4x^2 + b^4y^2}$ and mc^2z, and as ſ. BZ : 1 :: coſ. CZ : ſ. $AQ = \frac{c^2 z}{\sqrt{a^4x^2 + c^4z^2}}$, and coſ. $Aq = \frac{a^2 x}{\sqrt{a^4x^2 + c^4z^2}}$, $\frac{y}{\varepsilon} =$ coſ. BO, $\frac{\sqrt{x^2 + z^2}}{\varepsilon} =$ ſ. BO, $\frac{z}{\varepsilon} =$ coſ. CO, $\frac{\sqrt{x^2 + y^2}}{\varepsilon} =$ ſ. CO as ſ. BO : 1 :: coſ. CO : ſ. $QA = \frac{z}{\sqrt{x^2 + z^2}}$, coſ. $QA = \frac{x}{\sqrt{x^2 + z^2}}$, and the arch $Qq = AQ - Aq =$ the meaſure of the angle OBZ, and coſ. $Qq =$

a^2

$\frac{a^2x^2+c^2z^2}{\sqrt{x^2+z^2}\sqrt{a^4x^2+c^4z^2}}$, hence cos. OZ = s. BO × s. BZ × cos. Qq + cos. BO × cos. BZ = $\frac{\sqrt{x^2+z^2}}{\varkappa}$ × $m\sqrt{a^4x^2+c^4z^2}$ × cos. $Qq + \frac{y}{\varkappa}$ × $mb^2y = \frac{m}{\varkappa} \times (a^2x^2+c^2z^2+b^2y^2) = \frac{a^2\mathfrak{A}^2+c^2\mathfrak{C}^2}{\varkappa\sqrt{a^4\mathfrak{A}^2+c^4\mathfrak{C}^2}}$, the very same as before, proving the truth of the supposition. And by the nature of the motion as rad. = 1 : $\varkappa$ :: cos. OZ = s. OY : $\frac{a^2\mathfrak{A}^2+c^2\mathfrak{C}^2}{\sqrt{a^4\mathfrak{A}^2+c^4\mathfrak{C}^2}}$ = the velocity of the moving spherical surface at Y. Now, it does not appear, that there is any one point upon the varying great circle ZOY, which (in general) continues always the same or invariable upon the moving spherical surface, to find therefore the path of O about Z in absolute space, it is necessary to consider, that the point O, wheresoever upon the spherical surface it is found, can have but one proper direction of motion and velocity with which it shifts its place; those therefore in absolute space, and on the moving surface, must necessarily be the same, and consequently the two tracks, *viz.* that on the moving surface, and that on the fixed one or about Z in absolute space, must in all cases necessarily touch and roll. The fluxion of the track of O being $\sqrt{\dot\beta^2+\dot\gamma^2+\dot\delta^2} = \frac{\dot\varkappa}{\varkappa^3}\sqrt{\frac{\overline{\mathfrak{A}^2-BCe^2}|^2}{\beta^2}+\frac{\overline{\mathfrak{C}^2-ABe^2}|^2}{\delta^2}+\frac{A^2C^2e^4}{\gamma^2}}$, and the velocity along it $= \frac{\beta\gamma\delta}{ABC}\sqrt{\left(\frac{\overline{\mathfrak{A}^2-BCe^2}|^2}{\varkappa^2\beta^2}+\frac{\overline{\mathfrak{C}^2-ABe^2}|^2}{\varkappa^2\delta^2}+\frac{A^2C^2e^4}{\varkappa^2\gamma^2}\right)}$ because $\dot\gamma^2 = \frac{A^2C^2e^4\dot\varkappa^2}{\varkappa^6\gamma^2}$, $\dot\delta^2 = \frac{\overline{\mathfrak{C}^2-ABe^2}|^2\times\dot\varkappa^2}{\varkappa^6\delta^2}$, $\dot\beta^2 = \frac{\overline{\mathfrak{A}^2-BCe^2}|^2\times\dot\varkappa^2}{\varkappa^6\beta^2}$, $\gamma^2 = \frac{AC}{\varkappa^2}\times\overline{e^2-\varkappa^2}$, $\delta^2 = \frac{\mathfrak{C}^2-AB\times\overline{e^2-\varkappa^2}}{\varkappa^2}$, $\beta^2 = \frac{\mathfrak{A}^2-BC\times\overline{e^2-\varkappa^2}}{\varkappa^2}$; hence

$$\frac{(\mathfrak{A}^2-BC\times\overline{e^2+\varkappa^2})\times(\mathfrak{A}^2-BC\times\overline{e^2-\varkappa^2})}{\varkappa^2\beta^2} = \frac{\overline{\mathfrak{A}^2-BCe^2}|^2-B^2C^2\varkappa^4}{\varkappa^2\beta^2} = \mathfrak{A}^2 - BC$$

$\times \overline{e^2 + ȣ^2}$ in like manner $\frac{\overline{\mathfrak{C}^2 - ABe^2}|^2 - A^2B^2ȣ^4}{ȣ^2\delta^2} = \mathfrak{C}^2 - AB \times \overline{e^2 + ȣ^2}$, and $\frac{A^2C^2 \times \overline{e^4 - ȣ^4}}{ȣ^2\gamma^2} = AC \times \overline{e^2 + ȣ^2}$, hence the velocity $\frac{\sqrt{\dot{\beta}^2 + \dot{\gamma}^2 + \dot{\delta}^2}}{\dot{t}} = \frac{\beta\gamma\delta}{ABC}\sqrt{\left(\frac{B^2C^2ȣ^2}{\beta^2} + \frac{A^2B^2ȣ^2}{\delta^2} + \frac{A^2C^2ȣ^2}{\gamma^2} + \mathfrak{A}^2 - BCe^2 - BCȣ^2 + \mathfrak{C}^2 - AB \times \overline{e^2 + ȣ^2} + AC \times \overline{e^2 + ȣ^2}\right)} = \frac{\beta\gamma\delta}{ABC}\sqrt{\left(\frac{B^2C^2ȣ^2}{\beta^2} + \frac{A^2B^2ȣ^2}{\delta^2} + \frac{A^2C^2ȣ^2}{\gamma^2} - ȣ^2\right)}$ (becauſe $1 + AC - BC - AB = 0$, and $\mathfrak{A}^2 + \mathfrak{C}^2 = e^2$) which may be farther reduced to $\frac{ey}{ȣ^2}\sqrt{\left(\frac{\mathfrak{C}^2}{A^2BC} + \frac{\mathfrak{A}^2}{ABC^2} - \frac{e^2}{AC} - \frac{\overline{A + C} \times \mathfrak{A}^2\mathfrak{C}^2}{A^2BC^2e^2} + \frac{\mathfrak{A}^2\mathfrak{C}^2}{B^2y^2}\right)}$ = the velo- with which the momentary pole O ſhifts its place along its proper track; but it ſhifts its place in a direction perpendicular to the great circle ZO at O with a velocity whoſe ſquare is equal to the ſquare of that laſt found *minus* the ſquare of $\frac{\dot{ZO}}{\dot{t}}$ which is the velocity along $ZO = \frac{\dot{ȣ}}{\dot{t}ȣ^2 \times ſ.\,ZO} \times ȣ \times coſ.ZO = \frac{ȣ\beta\gamma\delta}{ABC \times tang.\,ZO}$, hence then the velocity perpendicular to ZO at O $= \frac{ȣ\beta\gamma\delta}{ABC}\sqrt{\left(\frac{B^2C^2}{\beta^2} + \frac{A^2B^2}{\delta^2} + \frac{A^2C^2}{\gamma^2} - 1 - \frac{1}{tang.\,ZO^2}\right)}$ this drawn into $\dot{t}$ gives $\frac{\dot{ȣ}}{ȣ}\sqrt{\left(\frac{B^2C^2}{\beta^2} + \frac{A^2B^2}{\delta^2} + \frac{A^2C^2}{\gamma^2} - \frac{1}{ſ.\,ZO^2}\right)}$ = the elementary ſpace perpendicular to ZO; hence the angular velocity with which O ſhifts its place about Z in abſolute ſpace $= \frac{ȣ\beta\gamma\delta}{ABC \times ſ.\,ZO}\sqrt{\left(\frac{B^2C^2}{\beta^2} + \frac{A^2B^2}{\delta^2} + \frac{A^2C^2}{\gamma^2} - \frac{1}{ſ.\,ZO^2}\right)}$, and the elementary ſpace divided by ſ. ZO gives the meaſure of the elementary angle, and the track of O in abſolute ſpace may hence, *conceſſis quadraturis*, be conſtructed by points. But this is unneceſſary after the path of one of the angles C of the octant has been found; ſince the track of O is thence given by the projection of points *ad libitum* of the now known triangle ZOC.

4 C 2 Hence

Hence then we collect, that the point Z is fuch that the angular velocities at the points *q*, *r*, *s*, Y, in directions perpendicular to the great circles drawn through Z and the poles A, B, C, and O, meafured at 90° diftance from Z, are all conftant quantities in all poffible cafes, notwithftanding the irregularity of the body's motion, which is a property very remarkable.

4. If $\frac{1}{A}$ here be $= 0$, $= \frac{b^2 - c^2}{a^2}$, and $b^2 = c^2$, or the two lefs *momenta* of *inertia* are equal, which is the cafe of a fquare prifm, cylinder, fpheroid, or other folid of revolution; then $\varepsilon^2 = e^2$ conftant, $B = C$, $\delta^2 = \frac{\mathfrak{C}^2}{e^2} - \gamma^2$, $\beta^2 = n^2 = \frac{\mathfrak{A}^2}{e^2}$ conftant, $e^2\beta^2 = x^2 = \mathfrak{A}^2$, $e^2\delta^2 = z^2 = \mathfrak{C}^2 - y^2$, $\dot{t} = -\frac{B\dot{y}}{\mathfrak{C}\sqrt{\mathfrak{C}^2 - y^2}} = \frac{B\dot{z}}{\mathfrak{A}\sqrt{\mathfrak{C}^2 - z^2}} = \frac{B\dot{\delta}}{\mathfrak{A}\sqrt{\frac{\mathfrak{C}^2}{e^2} - \delta^2}} = \frac{Be}{\mathfrak{A}\mathfrak{C}} \times \frac{\mathfrak{C}\dot{\delta}}{e\sqrt{\frac{\mathfrak{C}^2}{e^2} - \delta^2}}$, &c. as in the particular cafe confidered in the 4th and 5th propofitions, the A there being = B here. And hence the velocity above of O in its track $= \frac{\mathfrak{A}\mathfrak{C}}{Be} = \frac{eb\beta}{B}$ as there found. Cof. OZ = a conftant quantity $= \frac{a^2\mathfrak{A}^2 + c^2\mathfrak{C}^2}{e\sqrt{a^4\mathfrak{A}^2 + c^4\mathfrak{C}^2}} = \frac{a^2\beta^2 + c^2b^2}{\sqrt{a^4\beta^2 + c^4b^2}}$, and f. OZ $= \frac{b^2\mathfrak{A}\mathfrak{C}}{Be\sqrt{a^4\mathfrak{A}^2 + c^4\mathfrak{C}^2}} = \frac{b^2\beta b}{B\sqrt{a^4\beta^2 + c^4b^2}} = \frac{\overline{a^2 - c^2} \times \beta b}{\sqrt{a^4\beta^2 + c^4b^2}} = \frac{\beta b}{\sqrt{B^2 + 2B\beta^2 + \beta^2}}$, as there found, &c. And therefore, when *b* is very fmall, this is much fmaller, being then nearly $= \frac{b}{B+1}$, which in the cafe of the *earth* is nearly $= \frac{b}{232}$, and therefore infenfible. For on the hypothefis that $b = 9''$, this quantity, or half the diurnal nutation will be lefs than the $\frac{1}{25}$th part of a fecond, and the whole *diurnal nutation* lefs than the $5'''$. Indeed the $\frac{1}{13}$th part

part of a ſecond muſt be very near the true quantity; for, though the earth's figure may not be preciſely that of a ſpheroid, it cannot differ from it ſo much as to make any ſenſible alteration in this, eſpecially now it appears from the foregoing general ſolution, that the angular velocity about the axis whoſe pole is Z is always uniform and conſtant, let the figure of the revolving body be what it will. Neither can the progreſſive or annual motion cauſe any alteration, becauſe it cannot at all affect the rotatory or diurnal one.

5. The remarkable property mentioned at the end of the 3d of theſe general ſcholia, may be more particularly expreſſed thus: as the ſ. $Zq = mb^2y : 1 :: y : \frac{1}{mv^2} = \frac{\sqrt{a^4\mathfrak{A}^2 + c^4\mathfrak{C}^2}}{b^2}$ = the angular velocity at q about the axis whoſe pole is Z; in like manner, the angular velocity at r (fig. 4.) about the ſame axis $= \frac{\sqrt{a^4\mathfrak{A}^2 + c^4\mathfrak{C}^2}}{c^2}$, that at $s = \frac{\sqrt{a^4\mathfrak{A}^2 + c^4\mathfrak{C}^2}}{a^2}$, and that at $Y = v \text{ coſ.} OZ = \frac{a^2\mathfrak{A}^2 + c^2\mathfrak{C}^2}{\sqrt{a^4\mathfrak{A}^2 + c^4\mathfrak{C}^2}} = V = \sqrt{e^2 - \frac{\overline{a^2 - c^2}|^2 \times \mathfrak{A}^2\mathfrak{C}^2}{a^4\mathfrak{A}^2 + c^4\mathfrak{C}^2}}$, which, being the velocity of the moving ſpherical ſurface at every point of the great circle whoſe node is Y, and every point of that great circle being at the diſtance of 90° from Z, *the angular velocity of the body round the axis at reſt in abſolute ſpace whoſe pole is Z will be always equable, uniform, and conſtant, notwithſtanding the other oſcillating, vacillating motions of the body* : e being the greateſt angular velocity about the momentary axis.

This motion, then, is of the moſt ſimple and evident kind, and, together with that of the track under Z above determined, limits the whole compound motion under conſideration, all the others being only neceſſary conſequences of theſe; ſo fo that after all the pains beftowed upon the problem, the refult is as fimple as could be wifhed for; and the motion, though not quite fo regular, is as eafy to be conceived as that in the particular cafe of the folids of revolution. For the fpherical furface, concentric with the body, moves with an uniform and conftant angular velocity V about an axis IZ at reft in abfolute fpace, whilft the track WV′W′V upon that furface always paffes Z, the pole of that axis, with a velocity = $\sqrt{\frac{b^4\mathfrak{A}^2\mathfrak{C}^2}{B^2 \times \overline{a^4\mathfrak{A}^2 + c^4\mathfrak{C}^2}} - \frac{y^2}{AC}}$, which, though not conftant, recovers its firft value again and again in equal times, as the body revolves for ever.

6. I fhall only juft add, that if P, Q, and R, be any three external motive forces fuppofed to act upon the body in the directions of the three great circles BC, CA, and AB, then muft $\frac{P}{Ma^2} = \frac{\dot{x}}{\dot{t}} - \frac{b^2 - c^2}{a^2} \times yz$, $\frac{Q}{Mb^2} = \frac{\dot{y}}{\dot{t}} - \frac{c^2 - a^2}{b^2} \times zx$, and $\frac{R}{Mc^2} = \frac{\dot{z}}{\dot{t}} - \frac{a^2 - b^2}{c^2} \times xy$ exprefs the values of the external accelerating forces that act upon the body to alter its velocity about the three permanent axes of rotation. And when the relations of thofe external forces to the internal perturbating ones are given, a folution will hence be obtained to the more general problem, for determining the motion of the body, when, befides the perturbation arifing from the centrifugal force of its own particles, it is alfo acted upon by any external difturbing forces whatever. And, if P, Q, and R, be equal to, but in contrary directions to $Myz \times \overline{b^2 - c^2}$, $Mzx \times \overline{c^2 - a^2}$, and $Mxy \times \overline{a^2 - b^2}$, the perturbations vanifh, and then about whatever axis the body is firft impelled, it muft continue to revolve uniformly round it for ever.

Fig. 1.

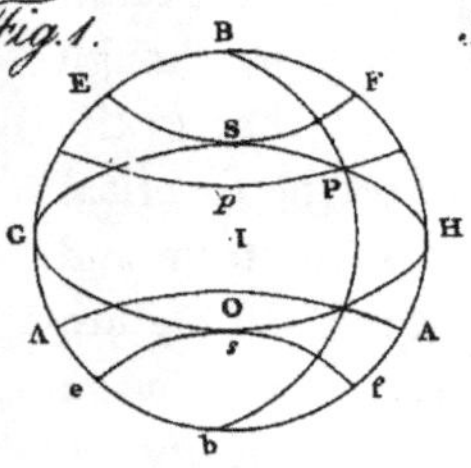

Fig. 2.

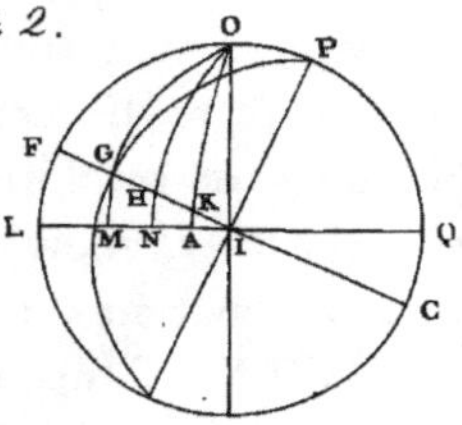

Fig. 3.

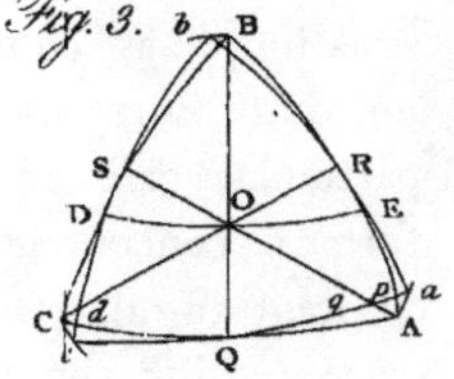

Fig

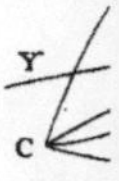

Fig. 5.

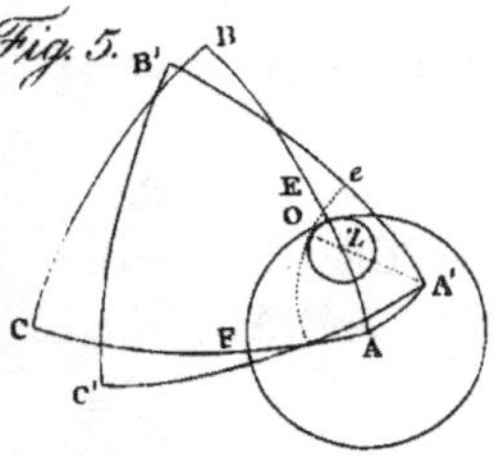

Fig. 6.

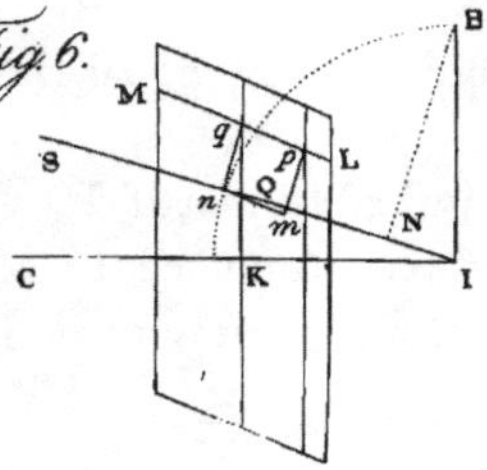

Fig. 7.

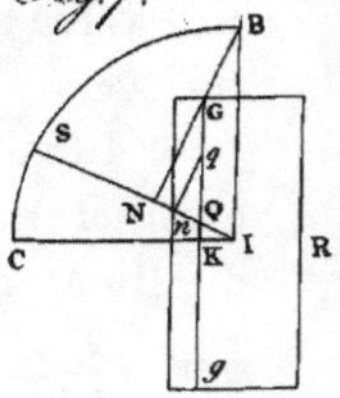

Fig. 8.

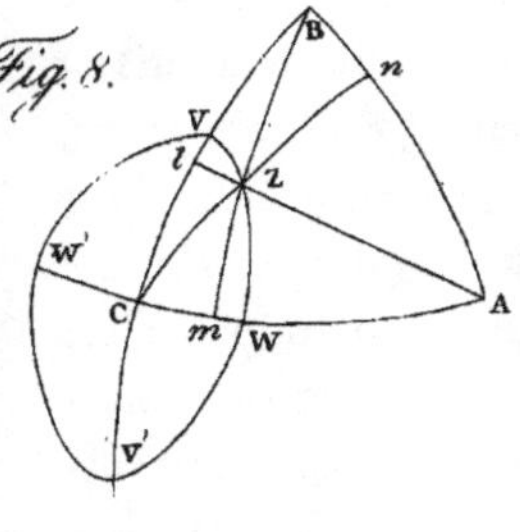

Fig. 9.

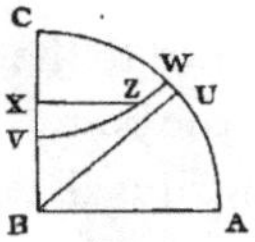

Fig. 10.

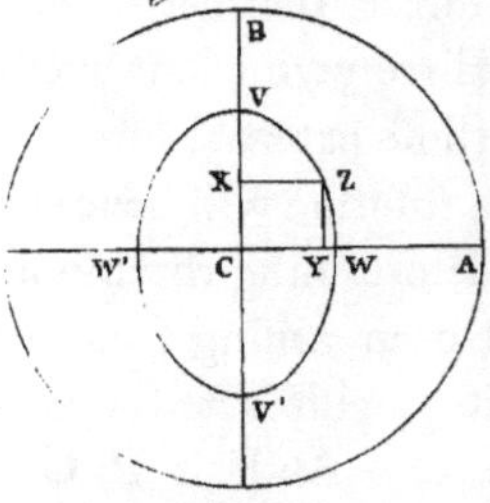

Fig. 4.

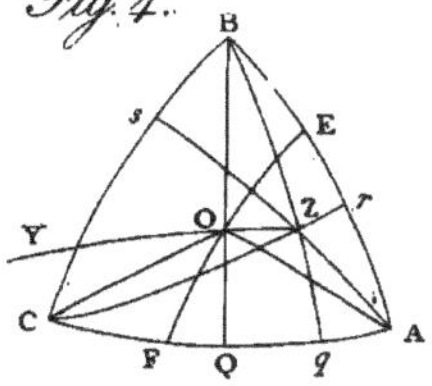

Fig. 12.

R

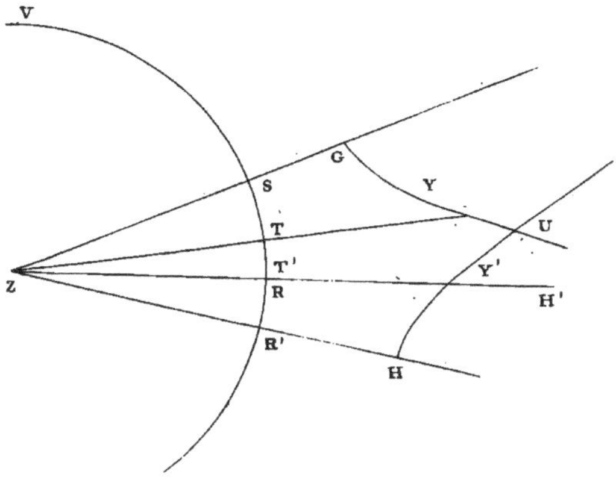

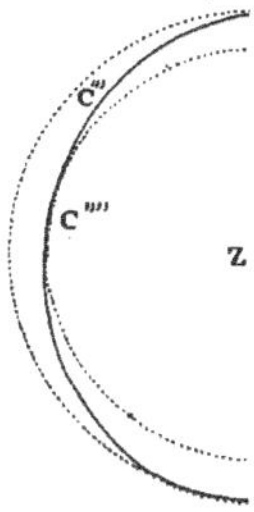

Fig. 11.

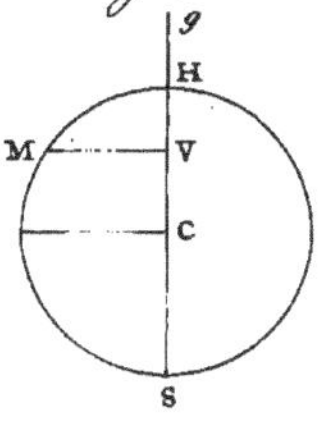

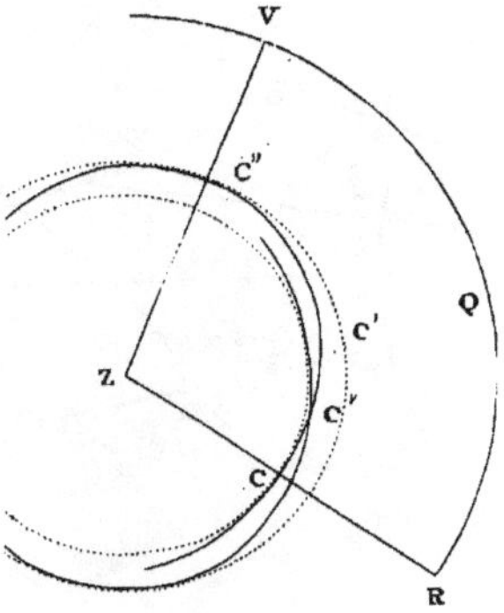

Basire sc.

Note referred to in page 519.

(C) Without any regard to the parallelopipedon, let the form of the body be what it will, if the *momenta* of *inertia* round the three permanent axes be repreſented by Maa, Mbb, and Mcc, the relative motive forces round thoſe axes will always be expreſſed by Ma^2x^2, Mb^2y^2, and Mc^2z^2, acting at the diſtance of radius therefrom. And then, in fig. 7., the centrifugal motive force acting along BI, being $= Mb^2y^2$, that acting along BN at N will, by the laws of central force, be $= Mb^2y^2 \times \frac{\mathrm{BI}}{\mathrm{BN}}$; and therefore the equivalent one, acting at S perpendicular to SI, will be $= Mb^2y^2 \times \frac{\mathrm{NI}}{\mathrm{BN}} = Mb^2y^2 \times \frac{x}{y} = Mb^2yx$ urging the point S towards B. In like manner it is found, that the centrifugal motive force Ma^2x^2 acting along CI produces one at S perpendicular to SI $= Ma^2xy$ urging it towards C; and the difference of theſe $= Ma^2xy - Mb^2xy$ muſt be the perturbating motive force at S, along the great circle BSC, as found by the other methods. And in the very ſame manner may thoſe in the other great circles bounding the octant be found.

The seven Satellites of Saturn as they appear and to be estimated the 18th of Octr at 7h 30' 55" 1789

Fig 1

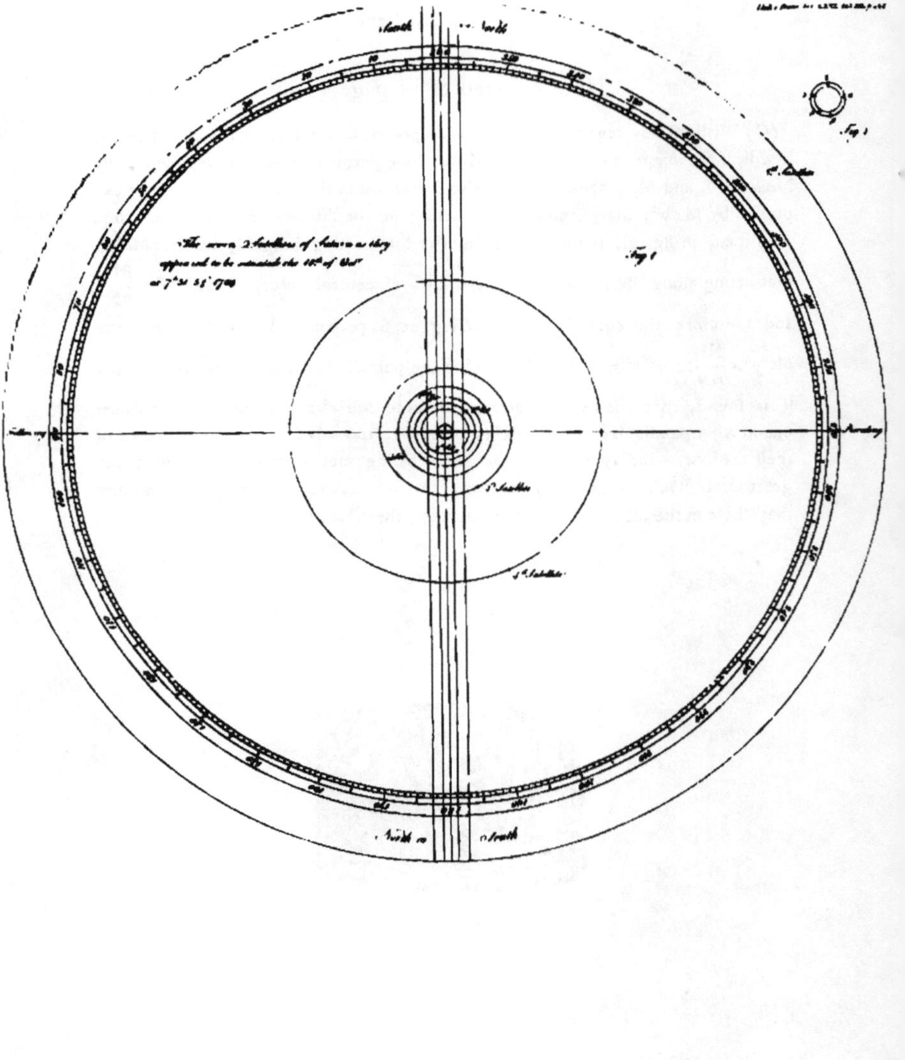

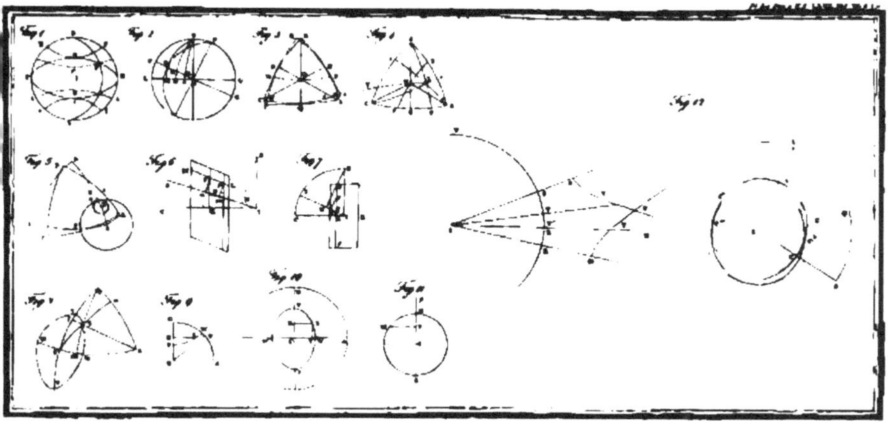

www.ingramcontent.com/pod-product-compliance
Lightning Source LLC
Chambersburg PA
CBHW022000190326
41519CB00010B/1344
* 9 7 8 3 3 3 7 1 0 2 3 3 3 *